EXPLORING CALCULUS WITH MAPLE®

Carmen Q. Artino
College of Saint Rose

Julian R. Kolod
College of Saint Rose

Benny Evans
Oklahoma State University

Jerry Johnson
University of Nevada, Reno

to accompany

CALCULUS

Deborah Hughes-Hallett
Harvard University

Andrew M. Gleason
Harvard University

et al.

John Wiley & Sons, Inc.
New York • Chichester • Brisbane • Toronto • Singapore

Copyright © 1994 by John Wiley & Sons, Inc.

This material may be reproduced for testing or instructional purposes by people using the text.

ISBN 0-471-09703-9

Printed in the United States of America

10 9 8 7 6 5 4 3 2 1

Table of Contents

Preface v

Chapter 1   A Library of Functions 1

       • Functions • Graphs • Power functions • Exponential functions • Logarithms • Inverses of functions • Trigonometric functions • Rational functions

Chapter 2   Key Concept: The Derivative 65

       • Velocity • Tangent lines • Derivatives • Limits

Chapter 3   Key Concept: The Definite Integral 97

       • Riemann sums • Definite integrals • The Fundamental Theorem • Area • Average value

Chapter 4   Short Cuts to Differentiation 123

       • Difference Quotients • Differentiation Rules • Implicit Differentiation

Chapter 5   Using the Derivative 137

       • Maxima and minima • Inflection points • Applications • Newton's method

Chapter 6   Reconstructing a Function from Its Derivative 169

       • Antiderivatives • Graphing antiderivatives

Chapter 7   The Integral 175

       • Approximating integrals • Improper integrals

Chapter 8   Using the Definite Integral 211

       • Applications • Arc length • Volume • Distributions

Chapter 9   Differential Equations 237

       • Families of solutions • Slope fields • Euler's method • Applications of first-order equations • Equilibrium solutions • Systems of equations • Second-order equations

Chapter 10   Approximations 275

   • Taylor polynomials • Interval of convergence • Fourier series

Appendix I   Maple V Release 3 Reference and Tutorials 297

Appendix II   Optional Files for Use in Maple V Release 3 323

Appendix III Common Maple Syntax 333

Index of Solved Problems 335

Index of Laboratory Exercises 337

PREFACE

This book was first developed in conjunction with the computer program *DERIVE*®. The present manual basically translates that book for use with Maple V. The *DERIVE* manual was written by Benny Evans and Jerry Johnson and the revision to Maple V was done by Carmen Q. Artino and Julian R. Kolod. Maple V and *DERIVE* are computer algebra systems that have enjoyed remarkable acceptance in the educational environment, especially during the current reform movement in mathematics education. When doing a translation such as this one, problems are always encountered owing to differences in the way *DERIVE* and Maple respond to a given input. However, in the majority of cases the translation was straightforward. In those instances where it was not, we have tried to maintain the same spirit as the original text.

Maple V is an extensive and complex computer algebra system that is being updated about every year and a half. Release 3 of the program is becoming widely available as this manual is being written. Given the fact that some users may still be using Releases 1 and 2, we have tried to make this manual as independent as possible of a particular release. For example, the majority of the illustrations are Postscript© renderings of Maple plot structures. Where illustrations of actual screens are presented, they depict images from the Windows© version of Release 3. In addition, Appendix I contains a tutorial that reflects Release 3.

This manual is an enrichment supplement to the Calculus text by Hughes-Hallett, Gleason, et al., which was the result of a major NSF initiative in Calculus reform called **The Calculus Consortium based at Harvard.** We will refer to it throughout the book as the **CCH Text** for short.

The purpose of this manual is to help students use the Maple program as a tool to explore Calculus in the spirit of the CCH Text, beyond the level of rote calculations and superficial exercises. Most of the problems are different from those which one would normally expect to do with nothing but pencil and paper.

The authors of this text have taught from the CCH Text using Maple and have found that they make a marvelous pair. Although this manual is written as a supplement to the CCH Text, it may be used with any calculus text as a source of stimulating problems.

In the spirit of the CCH Text, this manual is written for students – not instructors. One common thread running through it and the CCH Text is the admonition, "EXPLAIN

Maple V is copyright by Waterloo Maple Software.
DERIVE is copyright by Soft Warehouse, Inc.
Postscript is copyright by Adobe Systems.
Windows is a trademark of Microsoft, Inc.

YOUR ANSWER." Written explanations in clear, complete English sentences are assumed to be part of all assignments.

Key Features

- No prior knowledge of Maple is required. Appendix I summarizes important commands and will help novice users get started, but students with a little Maple experience should be able to begin Chapter 1 with no difficulty.

- The book is written with the Calculus reform spirit clearly in mind. Accordingly, the problems go beyond the level of rote calculations and "template" exercises.

- Optional code for creating graphics and automating some calculations is provided in Appendix II.

- Exercises are designed as laboratory sheets that students may detach and hand in.

- A handy reference list showing textbook and handwritten syntax as compared to correct Maple syntax is provided in Appendix III.

- Most of the problems have been class tested by Evans and Johnson and others.

Maple is very impressive, but our central purpose is to teach *mathematics*, not to show off hardware and software. We have tried to include enough Maple commands and suggestions in both the Solved Problems and in Appendix I so that the reader will not have to spend a great deal of time referring to *First Leaves: A Tutorial Introduction*, the basic Maple V user manual. Appendix I, however, is not intended as a substitute for it.

Structure of the Book

The chapters are designed to follow the CCH Text. Each chapter starts with a set of Solved Problems followed by a set of Laboratory Exercises. The Solved Problems are examples that provide a context for Maple instructions that one is likely to encounter in the exercises, but they also contain useful mathematical hints and lessons as well.

The Laboratory Exercises are sets of problems that students should be able to do upon completion of the appropriate Solved Problem. They are sufficiently complex that solving them without assistance from a computer or graphing calculator is not practical. Most of them invite the student to discuss their observations and findings. Each Solved Problem and each Laboratory Exercise contains a reference to the section of the CCH Text where it fits.

The Maple V Program

> *The novice who has never used Maple V before should begin by reading Sections 1-5 of Appendix I.*

Maple V is a comprehensive computer algebra system that runs on almost every major computer system and some minor ones, too.

Any sophisticated software takes some practice and experience to master, but we have used Maple in our courses for two years and are convinced that one of its strongest educational points is that it is easy for students to learn and use. It is also powerful enough that many scientific and engineering organizations use it in their daily work.

This manual was written using Maple V Release 3. If you are using an earlier version, you will notice some differences in the output that we have given. There also are some features in Release 3 that are not available in earlier releases. We have tried to point these out where they occur and to provide a method of accomplishing the same result.

Suggestions for Incorporating Maple: What Has Worked

- We are fortunate to have a classroom with 21 Amiga 3000 computers, one for each of 20 students and one for the instructor, and each is equipped with Maple V. This classroom is dedicated to the interactive teaching of mathematics.

- We normally spend the first class meeting of the semester introducing the students to Maple. We give them a reference sheet similar to what appears in Appendix I and use it as a guide for the first class meeting. After that, we introduce commands as needed. Our experience is that students quickly become comfortable with Maple and that only minimal help is needed as the course progresses.

- While Maple is used on a daily basis during class, we also give outside lab assignments that are similar to the lab exercises that appear in this manual. We encourage students to work together on these assignments, but we insist that they turn in their own work.

- We have found that some classroom discussion of the assigned labs in which Maple may be used is desirable. When a course is not being taught in the classroom, it is open for use by students. We have made ourselves available in the room at these times to give whatever assistance a student may need.

- Maple is also available in the other computer labs on campus. Our experience has been that students encounter little or no trouble using Maple on different computer systems.

Important Conventions

We have tried to follow several conventions in this manual.

1. Maple is not menu driven. Commands must be typed into Maple at a prompt, then the **Return** key or **Enter** key must be pressed. See Appendix I for the details. In general, whenever a special key is to be pressed, we will indicate this by boldfacing the name of the key.

2. Whenever we give a Maple command, it will appear in a different style type usually called typewriter or teletype style. For example, `plot` and `simplify` are Maple commands. Also, whenever appropriate, we will give Maple's response immediately following the command, and then add a horizontal line that represents the *separator* feature in Maple. For example, if we are asking Maple to add 2 and 3, we will write

```
>2 + 3;
```

5

A Word to Students

Maple has essentially automated most of the standard algebra calculations you normally encounter, just as scientific calculators have done with arithmetic. It will simplify complicated expressions, solve equations, and plot graphs. It also will do most of the standard calculations that arise in Calculus such as finding derivatives and integrals. But that doesn't mean that Calculus and Algebra are obsolete or unimportant. Even though calculators will do arithmetic, we still have to know what questions to ask, understand what the answers mean, and realize when an obvious error has been made. In the same way, we still have to understand the definitions, concepts, and processes that are involved in Algebra and Calculus so that we will know what to ask Maple to do, what its answers mean, and be able to detect errors. Maple only does the calculations; you must still do the thinking.

Always view any computer or calculator output critically. Be alert for answers that seem strange – you may have hit a wrong key, entered the wrong data, or made some other mistake. It is even possible that the program has a bug! If a problem asks for the cost of materials to make a shoe box and you get $123.28 or −$0.55 you should suspect that something is wrong.

Clear comunication is *at least* as important in mathematics as it is in other fields. You should always write your answers in complete, logical sentences. Re-read what you have written and ask yourself if it really makes sense.

Before you approach the computer, work through as much of the assigned problem as you can with pencil and paper, taking note of exactly where you think the computer will be required and for what purpose. You may be surprised at how little time you will actually have to spend in front of the computer if you follow this advice.

Carmen Q. Artino and Julian R. Kolod

June 1994

Chapter 1
A Library of Functions

Two of the features of Maple, plotting and equation solving, will be used extensively in this course. The necessary Maple instructions are included in the Solved Problems that follow, but you may find it helpful to refer to Appendix I for an overview of plotting and equation solving.

Solved Problem 1.1: Domains, ranges, and zeros of functions (CCH Text 1.1)

Plot the graph and find the domain, range, and zeros of each of the following functions.

(a) $f(x) = \dfrac{x^3 - 7x^2 - x + 7}{50}$

(b) $f(x) = x^4 - \sqrt{1-x}$

(c) $f(x) = \sqrt{1-x} + \sqrt{x-1} + 0.5$

Solution to (a): First, we define the function f,

```
>f:= x -> (x^3 - 7*x^2 - x + 7)/50;
```

and Maple responds with

$$f := x \to \frac{1}{50}x^3 - \frac{7}{50}x^2 - \frac{1}{50}x + \frac{7}{50}$$

We next plot f over the default domain using the `plot` command:

```
>plot(f(x), x);
```

(See Figure 1.1a).

We can see that there is one zero between 5 and 10, but it is difficult to determine whether there are any others. To remedy this situation, we plot f again, but this time we will use a different *viewing window*:

```
>plot(f(x), x = -4..8);
```

The result is Figure 1.1b.

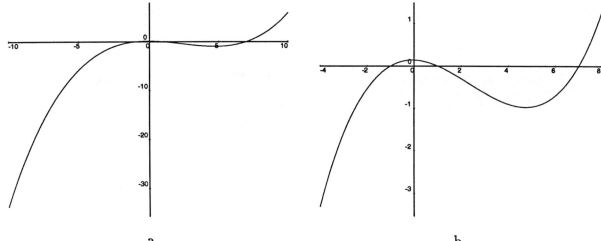

a.   b.
Figure 1.1: Different views of the graph of $\frac{x^3-7x^2-x+7}{50}$

It is now seen that f has three zeros and that the domain and range consist of all real numbers. We next ask Maple to solve the equation $f(x) = 0$ using the `solve` command.

```
>solve(f(x) = 0, x);
```

Maple responds with

$$1, 7, -1$$

Solution to (b): As in part (a), we first define f.

```
>f := x -> x^4 - sqrt(1 - x);
```
$$f := x \to x^4 - sqrt(1-x)$$

Plotting f over the default domain with the following command yields the result in Figure 1.2a:

```
>plot(f(x),x);
```

It isn't so easy to determine the domain and range from this graph. However, it is easy to see what the domain of $f(x) = x^4 - \sqrt{1-x}$ is without a computer! $1-x$ must be nonnegative to avoid taking the square root of a negative number. Hence the domain consists of all real numbers x that are less than or equal to 1. Using this, we zoom in on the graph

>plot(f(x), x = -1.5..1);

and obtain the picture in Figure 1.2b:

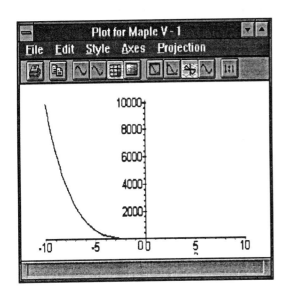 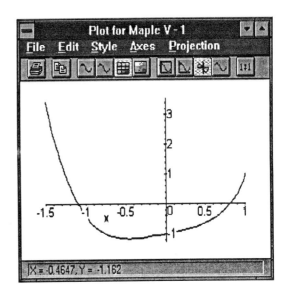

a. b.

Figure 1.2: Different views of the graph of $x^4 - \sqrt{1-x}$

The left side of the above graphs indicates that we can get functional values as large as we please, but we can't get smaller than the low point on the graph. By pointing and clicking at the low point on the graph, we see that its coordinates are about $(-.05, -1.16)$. Thus the range of f is approximately $[-1.16, \infty)$.

To find the zeros, we observe from the graph that there is one between -1.5 and -1 and another between 0.5 and 1. If we try solving the equation $f(x) = 0$ for x with

>solve(f(x) = 0, x);

$$\text{RootOf}(-1 + \_Z + \_Z^8)$$

The above response is what you will see if you are using Release 2 or earlier. If you are using Release 3 of Maple V, the response you see will be more complex. Maple cannot find the exact answer and is begging the question with its response — it is telling us that the zeros are the roots of the eighth degree polynomial, $-1 + x + x^8$. In these situations we can resort to the `fsolve` command on each of the intervals that the roots appear to find the approximate answers.

```
>fsolve(f(x)=0, x, -1.5..-1);
```
$$-1.096981558$$

```
>fsolve(f(x) = 0, x, 0.5..1);
```
$$.8116523200$$

Solution to (c): This is tricky, but it has a serious point. We first define the function,

```
>f := sqrt(1 - x) + sqrt(x - 1) + 0.5;
```

$$f := x \to sqrt(1-x) + sqrt(x-1) + .5$$

This time we will attempt to plot f over the default domain,

```
>plot(f(x), x);
```

Warning in iris plot: empty plot

What's going on here? Let's do some thinking. For $\sqrt{1-x}$ to be defined, we must have $x \leq 1$. For $\sqrt{x-1}$ to be defined we must have $x \geq 1$. Therefore, the domain of $\sqrt{1-x} + \sqrt{x-1} + .5$ consists of a single number, 1, and consequently the range consists of a single number, 0.5. There are, of course, no zeros of this function.

Remark: If you are using a version of Maple V earlier than Release 3, you will obtain a graph that is empty; that is, nothing is depicted except the axes. The graph doesn't always give complete information about the function. You may have to adjust the domain and/or range. Let's do this now. Try plotting the function over the interval $[0, 2]$, then $[0.5, 1.5]$, and lastly over $[0.75, 1.25]$ and observing the result in each case.

The steps used in these problems to make graphs and to solve equations are so important that they are worth summarizing once more.

PLOTTING GRAPHS

1. Define the expression or function you wish to plot.

2. Use Maple's `plot` command to plot the graph of the function or expression.

3. Zoom in or out to study the behavior of the graph.

4. Refer to Appendix 1 for further information on plotting.

APPROXIMATE SOLUTIONS OF EQUATIONS

For equations that Maple cannot solve exactly with the `solve` command, proceed as follows:

1. Write your equation as $f(x) = 0$ and plot the graph of $f(x)$.

2. Use the graph to isolate a root of the equation in a small interval, (a, b).

3. Use

   ```
   >fsolve(f(x) = 0, x, a..b);
   ```

   with the endpoints you found in step 2 for a and b. (If the function happens to be a polynomial, you may omit `a..b`; Maple will find all the real zeros in this case.)

4. Repeat steps 2 and 3 for the remaining roots.

5. Refer to Appendix 1 for further information on solving equations.

Laboratory Exercise 1.1

Zeros, Domains, and Ranges (CCH Text 1.1)

Name _____   Due Date _____

Plot the following functions. Find the zeros, domains, and ranges of each. Explain how you obtained your answers.

1. $f(x) = 3 + 15x^2 - x^4$

2. $g(x) = \sqrt{2x-1} + \sqrt{x-8} - x$

3. $h(x) = x^5 - x^3 - 900x - 901$. (You may need to alter the viewing window to get a good picture. What plotting intervals did you use?)

4. $i(x) = 0.25 + \sqrt{6-x} - \sqrt{x-6}$

5. $j(x) = \sqrt{-2x^4 + 9x^3 - 6x^2 - 11x + 6}$. (*Hint*: Factor the radicand.)

Solved Problem 1.2: Testing exponential data (CCH Text 1.3)

An experiment yields the following data:

$$(1, 0.075), \quad (3, 0.36), \quad (5, 1.75), \quad (7, 8.48), \quad (9, 41.05)$$

(a) Show that the data can be approximately fit by an exponential function ($y = ab^x$).

(b) Find the specific formula that fits the first two data points.

(c) Plot the data points and the exponential function on the same axes to confirm your findings.

Solution to (a): The basic principle is that an exponential function is one for which ratios of successive function values, with equal changes in the variable, are the same. Let's test the data by computing these ratios:

$$\frac{0.36}{0.075} = 4.8, \quad \frac{1.75}{0.36} = 4.861, \quad \frac{8.48}{1.75} = 4.846, \quad \frac{41.05}{8.48} = 4.841.$$

These results suggest that the data are approximately exponential.

Solution to (b): Now that we know the data fit the formula $y = ab^x$, we must find the numbers a and b. We can assert that $0.075 = ab^1$ and $0.36 = ab^3$. Dividing the second equation by the first gives $\frac{0.36}{0.075} = \frac{ab^3}{ab}$. This simplifies to $4.8 = b^2$ so $b \approx 2.19$.

Use the fact that $0.075 = ab$ to obtain $a = \frac{0.075}{b} \approx \frac{0.075}{2.19} \approx 0.0342$. We therefore conclude that the data approximately fit the function $y = (0.0342)2.19^x$.

Solution to (c): We will use Maple's ability to plot several items in a set. We define the set just as we would in mathematics using braces: { and }. Note also that in Maple, points are entered using square brackets instead of parentheses. At a Maple prompt, enter

```
>s := {[1, 0.075], [3, 0.36], [5, 1.75], [7, 8.48], [9, 41.05]};
```

$$s := \{[1, .075], [3, .36], [5, 1.75], [7, 8.48], [9, 41.05]\}$$

Now we can plot the points on coordinate axes by entering

```
>plot(s, style = POINT);
```

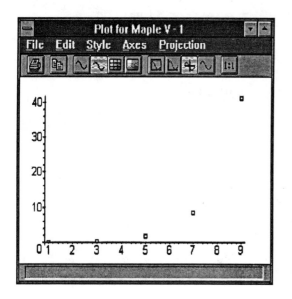
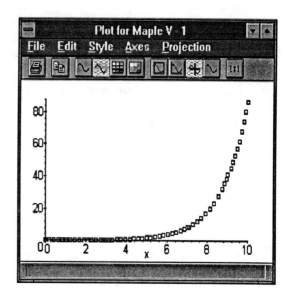

a. b.

Figure 1.3: Plot of Data and $.0342(2.19)^x$

Notice that the word POINT is spelled in all capital letters in this command. It is a syntax error to spell it otherwise. The result appears in Figure 1.3a. Now we will plot the points together with the function $y = (0.0342)2.19^x$ on the same axes. To do this we must insert the function into the set s. This is accomplished by forming the union of the set s with the set, $\{y\}$:

```
>y:=(0.0342)*2.19^x;
```
$$y := .0342\, 2.19^x$$

```
>s := s union {y};
```
$$s := \{.0342\, 2.19^x,\ [1,\ .075],\ [3,\ .36],\ [5,\ 1.75],\ [7,\ 8.48],\ [9,\ 41.05]\}$$

If we plot s over the domain $[0, 10]$, we will get the curve but not the points. In order to obtain both, we use an additional option to the `plot` command:

```
>plot(s, x = 0..10, style = POINT);
```

The result is shown in Figure 1.3b. Note that Maple has plotted the function $y = (0.0342)2.19^x$ as a series of points also.

Because the curve fits the data so well, it is difficult, if not impossible, to distinguish from the graph which points belong to the function and which points are the data points. (Ironically, the first release of Maple V would have plotted both the points and the function without resorting to the **style = POINT** option).

Laboratory Exercise 1.2

Fitting Exponential Data (CCH Text 1.3)

Name _____   Due Date _____

An experiment yields the following data.

$$(-2.5, 5.6),\ (-0.5, 2.75),\ (1.5, 1.35),\ (3.5, 0.66),\ (5.5, 0.32).$$

1. Show that the data can be approximately fit by an exponential function, $y = ab^x$.

2. Find the specific formula that approximately fits the data.

3. Plot the data points and the exponential function to confirm your findings. Be sure your plot shows all five data points.

Laboratory Exercise 1.3

U.S. Census Data (CCH Text 1.3)

Name _____   Due Date _____

This exercise refers to the table of census data in problem 16 of Section 2.4 of the CCH Text. In order to make a more readable plot, you should modify these data by taking the year 1790 as zero, 1800 as 10, and so on. Thus the first two points will be $(0, 3.9)$ and $(10, 5.3)$, and the last one will be $(200, 226.5)$.

1. Find the specific exponential formula, $y = ab^x$, that fits the first two data points. Plot the exponential function and all 21 of the modified data points together. Be sure your plot shows all the data points.

2. Estimate the first decade during which the curve and the data diverge. What historical event may have influenced this divergence?

3. If the population continued to grow at the same rate as it did during the period from 1790 to 1800, what would it have been in 1980? What would it be today?

4. Find the specific exponential formula, $y = ab^x$, that fits the decade 1890 to 1900. Plot this exponential function and all 21 of the modified data points together.

5. What are the two decades during the twentieth century where the data and the curve in Part 4 first diverge? What historical events may have influenced this divergence?

6. If the population continued to grow at the same rate as it did during the decade 1890 to 1900, what would it have been in 1980? What would it be today?

Solved Problem 1.3: Powers versus exponentials (CCH Text 1.4)

Plot the graphs of x^6 and 6^x. Determine the points where they intersect. For what values of x is $6^x > x^6$?

Solution: We first define the functions in Maple:

```
>f := x -> x^6;
```
$$f := x \to x^6$$

```
>g := x -> 6^x;
```
$$g := x \to 6^x$$

Since exponential functions grow quite rapidly, the default domain for `plot` would appear to be quite large (what's the value of 6^{10}?). Therefore, we restrict both the domain and range for the plot. See Figure 1.4a. (Which graph goes with which function?)

```
>plot({f(x), g(x)}, x = -2..2, y = 0..5);
```

We can see that the graphs cross between $x = -1$ and $x = 0$. If we point and click at the intersection point, we read the coordinates as $(-0.7891, 0.2611)$. Note that your results may be different depending on how close you click to the point of intersection. Following the technique presented in Part b of Solved Problem 1.1, we can get this more accurately. We use the `fsolve` command in the interval $(-1, 0)$:

```
>fsolve(f(x)=g(x), x, -1..0);
```

$$-.7898768569$$

We can see from Figure 1.4a that the two graphs seem to be coming back together on the positive x-axis. Where do they cross again? If we set the range to 0..30 we see a second crossing point in Figure 1.4b. This graph is the result of the following command:

```
>plot({f(x), g(x)}, x = -2..2, y = 0..30);
```

17

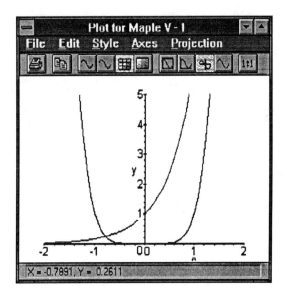 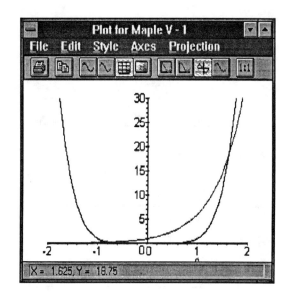

a. b.

Figure 1.4: Plot of x^6 and 6^x

This second crossing point lies between 1 and 2. (If we point and click, we see that it is at approximately 1.6.) Using `fsolve` again, we obtain the more accurate result,

>`fsolve(f(x) = g(x), x, 1..2);`

$$1.624243846$$

We do not need a computer to tell us that the graphs cross once more at $x = 6$. Thus $x^6 < 6^x$ on the interval $(-.7898768569, 1.624243846)$ and again on $(6, \infty)$.

Both of these graphs grow so rapidly that it is difficult to display all their interesting features on a single screen. Figure 1.5 is the result of executing the command,

>`plot({f(x), g(x)}, x = -7..7, y = 0..100000);`

It shows the crossing at $x = 6$. Notice that at this scale, the points of intersection near the origin cannot be seen.

This illustrates the fundamental concept that for bases greater than 1, exponential functions grow faster than power functions. That is, *if x is sufficiently large, then an* *exponential (a^x with $a > 1$) will eventually be larger than a power (x^b).* So 6^x is destined to eventually remain on top of x^6.

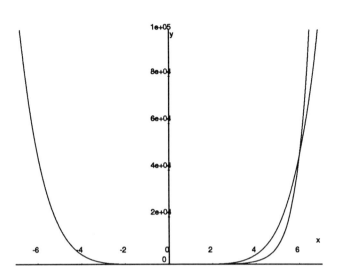

Figure 1.5: Expanded view of x^6 and 6^x

Laboratory Exercise 1.4

Powers versus Exponentials (CCH Text 1.4)

Name _____   Due Date _____

1. Plot the graphs of x^4 and 3^x. Determine the points where they intersect correct to ten significant digits (Maple's default value).

2. For what values of x is $3^x > x^4$?

3. Plot the graphs of x^5 and 2^x. Determine where they intersect correct to 10 significant digits.

4. Take the largest intersection point you found in Part 3 above and substitute it into $2^x - x^5$ for x. What do you get? $2^x - x^5$ *should* be zero there. Is it close to zero? Explain what's going on.

5. For what values of x is $2^x > x^5$?

Solved Problem 1.4: Inverses of functions (CCH Text 1.5)

Decide whether the following functions have inverses. For those that have inverses, (i) calculate $f^{-1}(4)$ and $f^{-1}(2)$, and (ii) plot both f and its inverse on the same coordinate axes.

(a) $\dfrac{x^4}{20} - 5x$

(b) $x^5 + 2x + 1$

Solution to (a): We first define the function f and `plot` it over the default domain; see Figure 1.6a.

```
>f := x -> x^4/20 - 5*x;
```
$$f := x \to \frac{1}{20}x^4 - 5x$$

```
>plot(f(x), x);
```

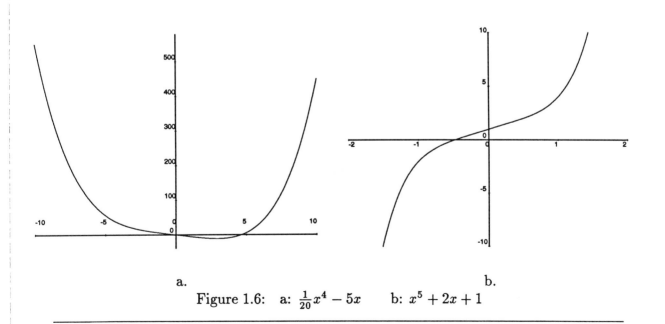

a. b.
Figure 1.6: a: $\frac{1}{20}x^4 - 5x$ b: $x^5 + 2x + 1$

We see immediately from the graph that f fails the horizontal line test and so does not have an inverse.

Solution to (b): We again begin by defining the function f in Maple.

```
>f := x -> x^5 + 2*x + 1;
```

$$f := x \to x^5 + 2x + 1$$

Increasing functions have inverses. So do decreasing ones. That's because the equation $f(x) = a$ cannot have more than one solution for x in these cases. (Explain why.) Our function, f, is increasing (see CCH Section 1.3 to recall the definition) as we can see by plotting. See Figure 1.6b for the result of the following command:

```
>plot(f(x), x = -2..2, y = -10..10);
```

We must be careful not to fall into a trap. The evidence provided by the graph in Figure 1.6b does not guarantee that it won't "wiggle" somewhere off the screen. We can verify that f is increasing by showing with algebra that if $x_1 < x_2$, then $f(x_1) < f(x_2)$. We leave it to the reader as an exercise to do this. (Show first that if $x_1 < x_2$; then $x_1^5 < x_2^5$ and $2x_1 < 2x_2$.) *We reiterate: this argument does not require a computer.*

Solution to (b) Part (i): It is not possible to find a general formula for the inverse of f, but we can calculate the special cases $f^{-1}(4)$ and $f^{-1}(2)$.

If $x = f^{-1}(4)$, then $f(x) = 4$. Thus, we need to solve the equation $x^5 + 2x + 1 = 4$. It is easy to see that $x = 1$ is a solution without a computer to help us, so we can say that $f^{-1}(4) = 1$. If $x = f^{-1}(2)$, then $f(x) = 2$ and we therefore need to solve the equation, $x^5 + 2x + 1 = 2$. For this, we can resort to the **fsolve** command. If we first plot f together with the line $y = 2$ on the same coordinate axes, we see that the graph of f and this line cross somewhere between 0 and 1. See Figure 1.7a. Thus, we use **fsolve** over this interval:

```
>fsolve(f(x) = 2, x, 0..1);
```
$$.4863890359$$

From this last exercise, we see that $f^{-1}(2) \approx .4863890359$.

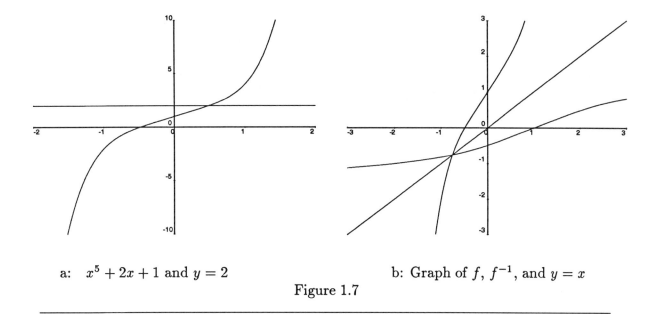

a: $x^5 + 2x + 1$ and $y = 2$

b: Graph of f, f^{-1}, and $y = x$

Figure 1.7

<u>Solution to (b) Part (ii):</u> Discussion: As mentioned above, it is not possible to find an explicit formula for $f^{-1}(x)$, but Maple can help us graph it nonetheless.

In CCH 1.5 we saw that one way to get the inverse of a function is to reverse the roles of x and y, and then solve the resulting equation for y. This is true because the graph of a function with domain D is the set of ordered pairs $(x, f(x))$ with x in D. Therefore, the graph of the inverse of $y = f(x)$ (if f has an inverse) ought to be the set of ordered pairs $(f(x), x)$ with x in D. This is the reason why the graphs of f and f^{-1} are symmetric about the line $y = x$, and it is also why we reverse the roles of x and y and attempt to solve the resulting equation for y. This can be a formidable task at times but is not really necessary when using Maple.

We will make use of a procedure that is part of Maple's very powerful plots package. This is a package of additional plotting routines that must be read into Maple's memory before they can be used. This is done with the command,

>with(plots);

Following the execution of this command, there will be a listing of the additional plotting procedures that are now available to us. If you inspect the list, you will see one called implicitplot and it is this plotting function that we will use to plot f, f^{-1}, and the line $y = x$. Simply put, we want to plot the three functions $y = f(x)$, $x = f(y)$, and $y = x$, and `implicitplot` makes it very easy to do. The result of

>`implicitplot({y = f(x), x = f(y), y = x}, x = -3..3, y = -3..3);`

appears in Figure 1.7b.

Discussion: Occasionally, as with our example, it may not be clear whether a graph is increasing. Maybe it will turn around off the screen and thus not be invertible after all. How do we know in general? For now we'll have to wait, but in Chapter 2 we will discuss the concept of the *derivative* which will help us with the problem.

Laboratory Exercise 1.5

The Inverse of a Function (CCH Text 1.5)

Name _____   Due Date _____

(a) Determine which of the following functions has an inverse and explain your answer.

(b) In each case, plot a graph that supports your assertion. If you assert that a graph fails the horizontal line test, plot a horizontal line that meets the graph at least twice. (You may have to alter the viewing window.)

(c) If f has an inverse, find $f^{-1}(2)$ and plot the graphs of f and f^{-1} together on the same axes.

1. $f(x) = x^3 - 3x^2 + 3x$

2. $f(x) = \dfrac{x^2}{20} - 5x$

3. $f(x) = x^3 + x - 2$

4. $f(x) = x^{\frac{11}{2}} + 2x^5$

5. $f(x) = x^5(\sqrt{x} + 2)2^{-x}$

Laboratory Exercise 1.6

Restricting the Domain (CCH Text 1.5)

Name _____   Due Date _____

If we restrict the domain of a function, we can make it pass the "horizontal line test" for invertibility. For example, $f(x) = x^2$ is not invertible, but $f(x) = x^2$ restricted to $[0, \infty)$ *is* invertible, and $f^{-1}(x) = \sqrt{x}$.

1. $g(x) = x^2$ on $(-\infty, 0]$ is invertible. What is $g^{-1}(x)$?

2. Show that $h(x) = x^3 + x^2 + 1$ is not invertible.

3. Find an interval of the form (a, ∞) on which $h(x) = x^3 + x^2 + 1$ is invertible, and estimate the smallest value of a for which this is so. Explain how you arrived at your estimate.

4. Plot the graphs of $h(x)$ and $h^{-1}(x)$ on the suitably restricted interval you found in Part 3.

5. Find two other intervals on which $h(x)$ above is invertible and plot the graphs of $h(x)$ and $h^{-1}(x)$ in these cases.

Laboratory Exercise 1.7

The Inverse of an Exponential Function (CCH Text 1.5)

Name _____   Due Date _____

1. Show that 10^x has an inverse. (Provide both graphical evidence and an algebraic argument.) For the remainder of this exercise we will refer to this inverse function as $L(x)$.

2. Find $L(10)$ and $L(20)$.

3. On the same axes graph 10^x, $L(x)$, and the line $y = x$. Explain the relationships among these three graphs.

4. Use the graph you have produced to explain why $L(x)$ is always less than x.

5. What is the domain and range of $L(x)$?

Important Information about MAPLE'S Syntax for Logarithms

Maple's syntax for $\log_b x$ is `log[b](x)`. Therefore, in order to calculate $\log_{10} 100$, enter `simplify(log[10](100));` at a Maple prompt and Maple will return the answer, 2.

In Maple, both `log(x)` and `ln(x)` refer to the *natural logarithm* or "log to the base e" ($e \approx 2.718$) as explained in CCH section 1.7. Be aware that this does not agree with the notation in your CCH text, which uses "log" to denote the logarithm to the base 10 and "ln" to denote the logarithm to the base e.

You should also be aware that when Maple simplifies any logarithm function, it first converts to natural logarithms, and the result may not be what you expect to see. For example, if you enter `log[10](x)` at a Maple prompt, you will see $\frac{\ln(x)}{\ln(10)}$. It is a fact that for any three positive numbers a, b and x, $\log_b x = \frac{\log_a x}{\log_a b}$. Maple uses this formula with e in place of a to define and calculate logarithms to bases other than e. By definition, $\log_b x = c$ means $b^c = x$. In other words, if you solve the equation $10^c = x$ for c, you should get $c = \log_{10} x$, but Maple will display the answer as $\frac{\ln(x)}{\ln(10)}$. Try it. At a Maple prompt, enter `solve(10^c = x, c);` and observe the result.

There is also a special way to enter the number e. If you just enter `e`, Maple will treat it like any other variable such as x, y, or a. In order to enter the special number $e \approx 2.718$, you must enter `E` (that's CAPITAL E). To see that this is correct, enter `evalf(E);` at a Maple prompt.

Solved Problem 1.5: Properties of logarithms from graphs (CCH Text 1.6)

(a) Plot the graph of $y = \log_{10}(2^x)$.

(b) Describe the graph in words and with a simple formula.

(c) Explain what this graph tells you about $\log_{10}(2^x)$.

Solution to (a): First, define the expression in Maple with

`>y := log[10](2^x);`

then when we plot this expression over the default domain

>plot(y, x);

we get the graph in Figure 1.8.

Solution to (b): The graph appears to be a straight line passing through the origin, very different from a logarithm curve. (Try enlarging the viewing window to verify this.) A straight line passing through the origin has an equation of the form $y = mx$. Therefore, there must be a number m so that $\log_{10}(2^x) = mx$. If we ask Maple to **solve** this equation, **simplify** the resulting solution, and then apply **evalf**, we obtain $m = .3010299957$. Thus the equation of the line is approximately $y = .3010299957x$.

Solution to (c): If we plotted this linear function, we would see that it overlays the graph of $\log_{10}(2^x)$. These graphs verify that $\log_{10}(2^x) = .3010299957x$. Since $\log_{10}(2^x) = x \log_{10}(2)$, we expect that $\log_{10}(2) = .3010299957$. Maple will readily verify this for you.

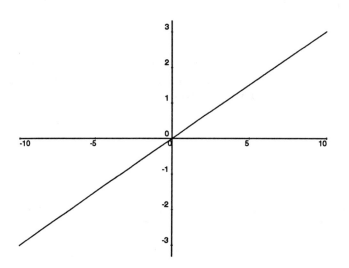

Figure 1.8:   Plot of $y = \log_{10}(2^x)$

Laboratory Exercise 1.8

Seeing Log Identities Graphically (CCH Text 1.6)

Name _____   Due Date _____

1. Plot the graph of $y = 10^{\log_{10} x}$. Describe this graph and explain what it tells you about $y = 10^{\log_{10} x}$.

2. Plot the graph of $y = 100^{\log_{10} x}$. How would you describe this graph?

3. Plot $y = x^2$ on the same axes as $y = 100^{\log_{10} x}$. When you compare the two graphs, what does it tell you?

4. Approximate $y = 100^{\log_{10} x}$ for various values of x using `evalf`. Discuss your observations and conclusions.

5. On the same axes plot $\log_2 x$, $\log_3 x$, $\log_4 x$, and $\log_5 x$. In general, if $b > 1$, discuss the effect of increasing b on the function $\log_b x$.

6. On the same axes graph $\log_2 x$ and $\log_{\frac{1}{2}} x$. What is the relationship between the two graphs? Conjecture how $\log_{\frac{1}{a}} x$ and $\log_a x$ are related in general. Try to prove your conjecture.

Solved Problem 1.6: Approximating the number e (CCH Text 1.7)

What is the smallest positive integer n so that $\left(1 + \dfrac{1}{n}\right)^n$ approximates e to two decimal places?

Solution: We know that e is approximately 2.71828, but let's see what Maple says anyway. Recall that e is represented in Maple as E. To get its decimal representation, we enter

```
>evalf(E);
```
$$2.718281828$$

We can make some quick computations of $\left(1 + \dfrac{1}{n}\right)^n$ using a "for" loop, a Maple construct for repeating calculations:

```
>f := n -> (1 + 1/n)^n;
```
$$f := n \rightarrow \left(1 + \dfrac{1}{n}\right)^n$$

```
>for n from 100 by 100 to 300 do n, evalf(f(n)); od;
        100, 2.704813829
        200, 2.711517123
        300, 2.713765158
```

The above table gives values of $f(n)$ for $n = 100, 200,$ and 300. (What happens in the above `for` loop if "by 100" is deleted?)

For an approximation to be correct to two decimal places, it must be within ± 0.005 of e. (Look at Appendix A of CCH in the part entitled Accuracy and Error, pages 660 – 662.) Therefore we want $\left(1 + \dfrac{1}{n}\right)^n$ to be between 2.713 and 2.723. We can see from the table above that $n = 200$ is too small. Note that for $n = 300$ that $\left(1 + \dfrac{1}{300}\right)^{300} \approx 2.71377$. So 300 works, but what is the *smallest* such n? More trial and error will be tedious and time consuming. Is there a better way?

Let's try graphing, but before doing so, note that the "for" loop has left n with the value 300. We must first set n back to a variable with

```
>n := 'n';
```
$$n := n$$

Now we can plot $e - (1 + 1/n)^n$ and look for the first place where the graph is below the horizontal line $y = 0.005$. Now let's think for a moment. We want to find the place where the graph is below the line $y = 0.005$ and we already know that this happens out near 300. So let's plot the graph of $g(n) = e - (1 + 1/n)^n$ together with the line $y = 0.005$ over the domain $200..300$.

```
>g := n -> E - (1 + 1/n)^n;
```
$$g := n \to E - \left(1 + \frac{1}{n}\right)^n$$

```
>plot({g(n), 0.005}, n = 200..300);
```

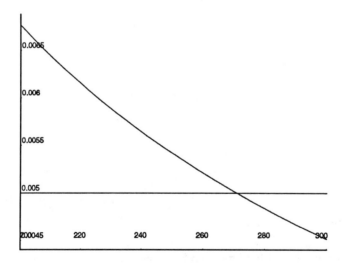

Figure 1.9: Plot of g and $y = 0.005$

The result of this plot may be seen in Figure 1.9. If we move the mouse pointer to the point where the graph of g crosses the line $y = 0.005$ and click, we will get an approximate value for n. This appears to be about 270 or so. To get a better value for n we compute the values of $g(270)$ and $g(271)$:

```
>evalf(g(270)); evalf(g(271));
```
$$.005016825$$
$$.004998375$$

From these results, we see that $n = 270$ is somewhat small so that our answer is $n = 271$.

A Second Solution: We want $g(n)$ to be just less than 0.005. Let's try using `fsolve` on the equation $g(n) = 0.005$. We tell Maple to look only in the interval $(200, 300)$ for a solution.

```
>fsolve(g(n) = 0.005, n, 200..300);
```
$$270.9116441$$

Since n is an integer, and the graph is decreasing, we choose $n = 271$.

Laboratory Exercise 1.9

Approximating e (CCH Text 1.8)

Name _____   Due Date _____

1. Plot $\left(1 + \dfrac{1}{n}\right)^n$.

2. What is the smallest positive integer n so that $\left(1 + \dfrac{1}{n}\right)^n$ approximates e to three decimal places? Explain how you arrived at your answer.

Laboratory Exercise 1.10

Seeing Log Identities Graphically II (CCH Text 1.7)

Name _____   Due Date _____

1. Plot $\ln(2x) - \ln x$. How would you describe the graph? (If you are confused by what you see on the screen, try altering the vertical range to 0 .. 1.)

2. What does the graph tell you about the relationship between $\ln(2x)$ and $\ln x$? Explain.

3. Plot $\ln(x^2) - 2\ln x$. How would you describe this graph? (Remember the parenthetical suggestion in Part 1.)

4. What does the graph tell you about the relationship between $\ln(x^2)$ and $2\ln x$?

5. Repeat all of the above parts with "3" in place of "2."

Laboratory Exercise 1.11

Growth Rates of Functions (CCH Text 1.7)

Name _____   Due Date _____

1. On the same axes graph $\ln x$, $x^{\frac{1}{2}}$, x, x^2, and e^x.

2. What is the relationship among these five functions for large values of x?

3. In general, what is the relationship among logarithmic functions, power functions, and exponential functions for large values of x?

Laboratory Exercise 1.12

A Graphical Look at Borrowing Money (CCH Text 1.8)

Name _____   Due Date _____

On the same axes, plot $\left(1 + \dfrac{0.1}{n}\right)^{nt}$ for $n = 1, 2, 4, 12$, and then plot $e^{0.1t}$. Explain in practical terms what the graphs tell you about borrowing money at 10% per year. Include in your explanation a discussion of the effects of time t and the number of compounding periods n. (Hint: Zoom in on the graph around $t = 20$.)

Laboratory Exercise 1.13

Shifting and Stretching (CCH Text 1.9)

Name _____   Due Date _____

Before beginning these exercises, define the function $f(x) = x^3 - x$ in Maple with

```
>f := x -> x^3 - x;
```

In this lab you will use Maple's `seq` command (short for *sequence*) to construct various sets of functions related to $f(x)$. For example, in the first problem, you will study the effect of adding a constant k to f; that is, you will be asked to plot $f(x) + k$ for $k = 0, 1, 2, 3, 4$. To do this, define a set S as follows:

```
>S := {seq(f(x) + k, k = 0..4)};
```

and Maple will produce the set,

$$S := \{x^3 - x, x^3 - x + 1, x^3 - x + 2, x^3 - x + 3, x^3 - x + 4\}.$$

The elements in this set may not appear in the order given here, but they will all be in S. Now you can `plot` S as usual. Note that you may have to alter the viewing window to get a good view of all the functions in S.

1. As described above, plot the set $S = \{f(x) + k \mid k = 0 \ldots 4\}$. Discuss in general how the graph of $f(x) + k$ varies with k.

2. Define and plot the set $S = \{f(x + k) \mid k = 0 \ldots 4\}$. Discuss how the graph of $f(x + k)$ varies with k.

3. Is it true that $f(x+k) = f(x) + k$? Explain.

4. Define and plot the set $S = \{f(x-k) \mid k = 0\ldots 4\}$. Discuss how the graph of $f(x-k)$ varies with k.

5. How do the graphs of $f(x-k)$ compare to the case $f(x+k)$?

6. Define and plot the set $S = \{kf(x) \mid k = 1\ldots 4\}$. Discuss how the graph of $kf(x)$ varies with k.

7. Define and plot the set $S = \{f(kx) \mid k = 1\ldots 4\}$. Discuss how the graph of $f(kx)$ varies with k.

8. Is it true that $f(kx) = kf(x)$? Explain.

Information About Trigonometric Functions

Maple's syntax for the six trigonometric functions is standard. Placing the prefix, "arc" before the corresponding function gives its inverse. For example, `arcsin(x)` is the inverse of $\sin(x)$. Trig functions in Maple always assume radian measure. Thus `sin(30)` is the sine of 30 radians, not 30 degrees.

You enter the number π in Maple using `Pi` (that's Capital P, lower case i). Try entering `evalf(Pi);` at a Maple prompt and observe the result.

Solved Problem 1.7: Periodicity of trigonometric functions (CCH Text 1.10)

Make separate plots of $\sin^2 x$ and $\sin(x^2)$. Are they periodic? What are their periods?

Solution: Define the first function in Maple as follows:

```
>f := x -> sin(x)^2;
```
$$f := x \to \sin(x)^2$$

That's $\sin(x)^2$, not $\sin^2 x$. What's going on? This simply is a peculiarity of Maple. As far as Maple is concerned, $\sin(x)^2$, $\sin^2 x$, and $(\sin x)^2$ are all the same, but $\sin(x^2)$ is different as we shall see.

Now let's plot it with

```
>plot(f(x), x = -3*Pi/2..3*Pi/2);
```

The resulting graph is given in Figure 1.10a and looks periodic. The humps appear to repeat every π units, so we guess that's the period. To check this, enter `f(x + Pi);` at a Maple prompt and notice that Maple returns $\sin(x)^2$ as the result. Thus Maple has verified our conjecture that f is periodic and has period π. (This, of course, is not a proof. To do that we would need to establish that $f(x) = f(x + \pi)$.) Notice that $\sin(x)$ has period 2π, but $\sin^2 x$ has period π.

Now let's turn to the second function and call it g:

```
>g := x -> sin(x^2);
```
$$g := x \to \sin(x^2)$$

The result of plotting $g(x)$ over the interval $[-3\pi/2, 3\pi/2]$ is seen in Figure 1.10b. (The difference between this and the graph of $\sin^2 x$ is quite obvious. Here the square is inside the parentheses, whereas in the first example it is outside.) As we go out the x-axis, the graph seems to oscillate faster and faster — the humps are narrower. To see this, try plotting g over the interval $[-2\pi, 2\pi]$. This does *not* look periodic, and it isn't.

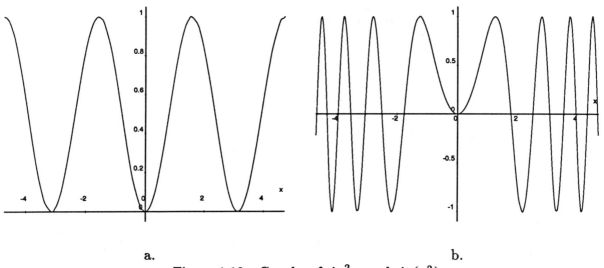

a. b.

Figure 1.10 Graphs of $\sin^2 x$ and $\sin(x^2)$

Laboratory Exercise 1.14

Equations Involving Trig Functions (CCH Text 1.10)

Name _____   Due Date _____

1. Plot $\cos(2x)$ and x^2.

2. Use the above plot and Maple's `fsolve` command to find <u>all</u> solutions of the equation $\cos(2x) = x^2$.

3. Explain how you can be sure you have found <u>all</u> solutions.

4. Use your answer in Part 2 to find all points of intersection of the graphs of $\cos(2x)$ and x^2.

Laboratory Exercise 1.15

Inverse Trig Functions (CCH Text 1.10)

Name _____   Due Date _____

1. Make plots of $\sin x$, $\arcsin x$, and $y = x$ on the same axes. Explain how the graphs are related.

2. Make plots of $\tan x$, $\arctan x$, and $y = x$ on the same axes. Explain how the graphs are related.

3. Make plots of $\cos x$, $\arccos x$, and $y = x$ on the same axes. Explain how the graphs are related.

Solved Problem 1.8: Rational functions (CCH Text 1.11)

Does $f(x) = \dfrac{2x^3 + x^2 - 15x + 7}{12x^2 - 3}$ have any horizontal asymptotes? How about $g(x) = \dfrac{2x^3 + x^2 - 15x + 7}{12x^3 - 3}$? Graph both functions together with their asymptotes.

Solution: First, we define the function f in Maple with

```
>f := x -> (2*x^3 + x^2 - 15*x + 7)/(12*x^2 - 3);
```

$$f := x \to \frac{2x^3 + x^2 - 15x + 7}{12x^2 - 3}$$

To test for horizontal asymptotes, we look at what happens as $x \to \infty$ and $x \to -\infty$. To do this, we use Maple's `limit` command as follows:

```
>limit(f(x), x = infinity);
```

$$\infty$$

You can re-execute this command with "infinity" replaced by "-infinity" and observe the result. These two results tell us that $f(x)$ has no horizontal asymptotes.

Now repeat the above two limit commands with $f(x)$ replaced by $g(x)$. In this case, we see that both limits are $\frac{1}{6}$, indicating a horizontal asymptote, $y = \frac{1}{6}$. In Figure 1.11, we have graphed both functions and added the horizontal asymptote for the second. Figure 1.11a was graphed with the command,

```
>plot(f(x), x = -6..6, y = -10..10);
```

Figure 1.11b was produced with the command,

```
>plot({g(x),1/6}, x = -6..6, y = -4..2);
```

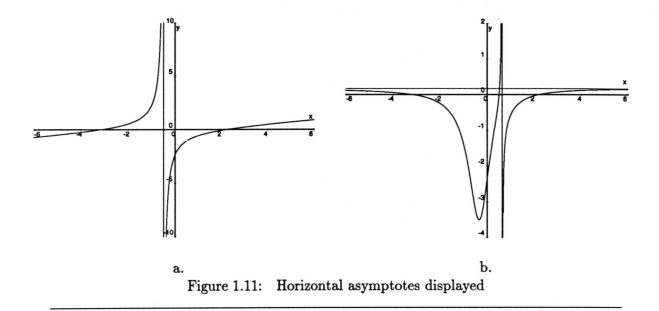

a. b.
Figure 1.11: Horizontal asymptotes displayed

Solved Problem 1.9: Vertical asymptotes (CCH Text 1.11)

Let $f(x) = \dfrac{2x^3 + x^2 - 15x + 7}{12x^2 - 3}$. Plot $f(x)$. How many vertical asymptotes do you see? How many do you expect to see? Explain. Add the graphs of the vertical asymptotes to the picture.

Solution: First, we define $f(x)$ as in Solved Problem 1.8 (see p. 59), and then we'll plot it as we did in that problem except that this time, we'll use the domain $[-4, 4]$ instead of $[-6, 6]$. This will provide us with a somewhat better view of the function near the asymptote. The result of the following command appears in Figure 1.12.

```
>plot(f(x), x = -4..4, y = -10..10, discont = true);
```

From this it appears that there is a vertical asymptote between $x = -1$ and $x = 0$. Note that we have used the option `discont = true` in this command. If we had not, Maple would have plotted what would appear to be a vertical asymptote there. It really isn't an asymptote; see Appendix I for more information on this point. The option discont = true can sometimes alleviate the problem but not always. In addition, this option is not available on releases earlier than Release 3. To get a truer picture, try the option style = POINT instead.

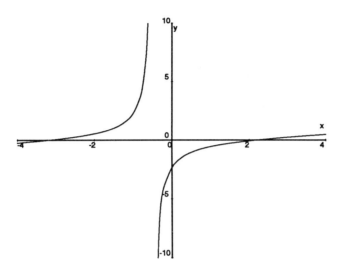

Figure 1.12: A phantom asymptote

To test for vertical asymptotes, we look for values of x where the denominator is zero. $12x^2 - 3 = 0$ for $x = \pm\frac{1}{2}$. Therefore, we are not surprised to see $x = -\frac{1}{2}$, which must be the one in the graph in Figure 1.12. But where is the one at $x = \frac{1}{2}$? To see why it is missing, we will factor the numerator and denominator.

Maple makes it easy for us to access the numerator and denominator of a rational expression because it provides us with two functions, numer and denom. numer(f(x)); will return the numerator of $f(x)$. So to factor the numerator, we enter

>factor(numer(f(x)));
$$(2x - 1)(x^2 + x - 7)$$

Factoring the denominator is just as easy:

>factor(denom(f(x)));
$$3(2x - 1)(2x + 1)$$

We can now see why there is no vertical asymptote at $x = \frac{1}{2}$. It is because there is a common factor of $2x - 1$ in both the numerator and denominator. When they are canceled out, we see that only the $2x + 1$ remains in the denominator. Hence, there is just one asymptote. If we simplify $f(x)$ with

>simplify(f(x));

Maple returns
$$\frac{1}{3} \frac{x^2 + x - 7}{2x + 1}$$

Remark: Since the original function is undefined at $x = \frac{1}{2}$, we would expect a "hole" in the graph. If you would like to see it, plot the graph of $f(x)$ with a very small domain and range, say x = 0.45..0.55 and y = -1.04..1. Then look very closely at the graph. You should be able to see it.

In order to plot the vertical asymptote at $x = -\frac{1}{2}$, we need to ask Maple to plot a straight line joining two points within the viewing window of the graph. Since our viewing window is defined by the domain $[-4, 4]$ and range $[-10, 10]$, it will be sufficient to plot the line segment joining the points $(-\frac{1}{2}, 10)$ and $(-\frac{1}{2}, -10)$. We can define points in Maple as a list containing an even number of items; in our case, this list will be [-1/2, -10, -1/2, 10] and we include this list in a set with $f(x)$. The result of the following command is given in Figure 1.13 shown below. This method can be used to graph vertical lines.

>plot({f(x), [-1/2, -10, -1/2, 10]}, x = -4..4, y = -10..10);

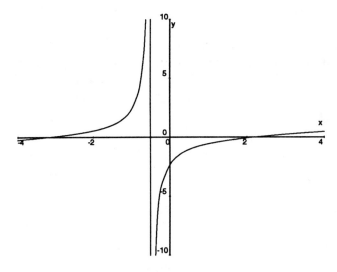

Figure 1.13: Vertical asymptote displayed

Laboratory Exercise 1.16

Asymptotes (CCH Text 1.11)

Name _____   Due Date _____

1. Make plots of the two functions in Solved Problem 1.8 that clearly show their "end" behavior, that is, the horizontal asymptotes or lack of them.

2. Explain how the graphs illustrate the results obtained in Solved Problem 1.8.

3. Find the vertical asymptotes of the second function in Solved Problem 1.8. Explain how you arrived at your answer.

4. Let $g(x) = \dfrac{3x^3 - 17x^2 + 16x - 4}{3x^3 - 2x^2 + 3x - 2}$. Find all the vertical asymptotes of $g(x)$.

5. Find all horizontal asymptotes of $g(x)$.

6. Plot $g(x)$ so the graph shows the aymptotes. (More than one plot may be necessary.)

7. Let $h(x) = \dfrac{3x^3 - 17x^2 + 16x - 4}{3x^3 - 2x^2 - 3x + 2}$. Find all the vertical asymptotes of $h(x)$.

8. Find all the horizontal asymptotes of $h(x)$.

9. Plot $h(x)$ so that the graph shows the asymptotes. (More than one plot may be necessary.)

Chapter 2
Key Concept: The Derivative

In this chapter, we investigate how average velocity over a time interval can be used to define and calculate instantaneous velocity. The idea is extended to explore average and instantaneous rates of change for general functions. As usual, graphing will be employed extensively, and you are advised to consult Appendix I for details on plotting with Maple.

Solved Problem 2.1: Calculating velocities (CCH Text 2.1)

Taking into account air resistance, a suitcase dropped from an airplane falls $968(e^{-0.18t} - 1) + 176t$ feet in t seconds.

(a) Find the average velocity of the suitcase over the time intervals $t = 1.99$ to $t = 2$ and $t = 2$ to $t = 2.01$. Use this to estimate the instantaneous velocity of the suitcase at time $t = 2$.

(b) Plot the graph of the average velocity from $t = 2$ to $t = 2 + h$ as a function of h and use it to estimate the instantaneous velocity at time $t = 2$.

(c) Find the instantaneous velocity of the suitcase at time $t = 2$ by calculating the appropriate limit. Compare this to the two estimates in Parts (a) and (b).

Solution to (a): First, define the distance function s as follows:

```
>s := t -> 968*(E^(-0.18*t) - 1) + 176*t;
```

$$s := t \to 968 E^{(-.18t)} - 968 + 176t$$

The average velocity over the time interval $[1.99, 2]$ is the distance traveled divided by the elapsed time.

$$\frac{s(2) - s(1.99)}{0.01}.$$

We can get an approximate value of this expression from Maple by entering

```
>evalf((s(2) - s(1.99))/0.01);
```

$$54.32741$$

The result is, of course, given in feet per second. Similarly, the average velocity over the time interval $[2, 2.01]$ is $\dfrac{s(2.01) - s(2)}{0.01} \approx 54.54622$ feet per second. Either of these is a reasonable estimate for the instantaneous velocity as is their average, 54.436815 feet per second. The CCH text customarily uses $\dfrac{s(2.01) - s(2)}{0.01} \approx 54.54622$ feet per second.

Solution to (b): The average velocity from $t = 2$ to $t = 2 + h$ is $\dfrac{s(2+h) - s(2)}{h}$. Since we will be interested in the values of this expression for h near 0, we plot it as follows:

```
>plot((s(2 + h) - s(2))/h, h = -1..1);
```

The graph is shown in Figure 2.1.

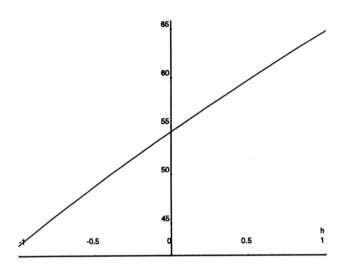

Figure 2.1: Average velocity for a falling suitcase

As mentioned above, we are interested in the value of this expression for h near 0. To get an approximation of the value of this expression for $h = 0$, we move the mouse pointer to where the graph crosses the vertical axis and click the mouse button. The approximate coordinates at the position of the click then appear on the graph. The second coordinate is the approximate value we seek and should be a reasonable estimate for the instantaneous velocity. <u>Remark</u>: There is actually a hole in this graph at $h = 0$. Look at the definition of the function and explain why.

Solution to (c): To compute the exact value of the instantaneous velocity of the falling suitcase at $t = 2$ we must compute $\lim\limits_{h \to 0} \dfrac{s(2+h) - s(2)}{h}$. If we compute this limit directly in Maple, we get the following result:

```
>limit((s(2 + h) - s(2))/h, h = 0);
```
$$undefined$$

Maple is telling us that in the present form of the expression, it cannot determine the limit we requested. However, if we **expand** the expression first, then take the limit, Maple will find the limit.

```
>expand((s(2 + h) - s(2))/h);
```

$$968\frac{E^{(-.18h)}}{hE^{.36}} + 176 - \frac{968}{hE^{.36}}$$

```
>limit(", h = 0);
```

$$54.43687688$$

Making a Table of Values

We can have Maple make a table of average velocities for Solved Problem 2.1 quite easily as follows. First, create a list in Maple with two entries, h and $\frac{s(2+h) - s(2)}{h}$.

```
>L := [h, evalf((s(2 + h) - s(2))/h)];
```

$$L := \left[h, \frac{968\ 2.718281828^{(-.36-.18h)} + 176\ h - 675.3506837}{h}\right]$$

Now we print the list L for various values of h in a **for** loop.

```
>for h from -0.05 by 0.01 to -0.01 do print(L); od;
```

Notice that we have deliberately skipped the value $h = 0$. If we hadn't, Maple would have terminated the loop with a "division by zero" error. The resulting table is given in Figure 2.2. You can print the values from 0.01 to 0.05 in a similar way.

[-.05, 53.88819800]
[-.04, 53.99819750]
[-.03, 54.10806667]
[-.02, 54.21780000]
[-.01, 54.32741000]

Figure 2.2: A table of average velocities

Laboratory Exercise 2.1

Average Velocity (CCH Text 2.1)

Name _____   Due Date _____

1. You drive for one minute at a constant velocity of 60 miles per hour. You then instantly slow down and drive for one more minute at a constant velocity of 40 miles per hour. Assuming you lost no time in slowing down, what is your average velocity for the two minutes?

2. You drive for one mile at a constant velocity of 60 miles per hour. You then instantly slow down and drive for one more mile at a constant velocity of 40 miles per hour. Assuming you lost no time in slowing down, what is your average velocity for the two miles? (Note: 50 miles per hour is not the right answer. If that is what you got, think once more about the definition of average velocity.)

3. You drive for one minute at a constant velocity of 30 miles per hour. You then want to instantly speed up and drive for another minute so that your average velocity for the two minutes is 60 miles per hour. Assuming you lost no time in speeding up, what velocity must you drive for the second minute?

4. You drive for one mile at a constant velocity of 30 miles per hour. You then want to instantly speed up and drive for another mile so that your average velocity for the two miles is 60 miles per hour. Assuming you lost no time in speeding up, what velocity must you drive for the second mile? (Be careful.)

Laboratory Exercise 2.2

Falling with a Parachute (CCH Text 2.1)

Name _____   Due Date _____

If a man jumps from an airplane with a parachute, he will fall

$$s(t) = 12.5(e^{-1.6t} - 1) + 20t$$

feet in t seconds. Answer the following questions about the parachutist. (Be sure to present your answers in appropriate units.)

1. Find the parachutist's average velocity over the time intervals $[0.99, 1]$ and $[1, 1.01]$. Estimate the instantaneous velocity at $t = 1$. Explain how you arrived at your answers.

2. Plot the graph of the parachutist's average velocity from $t = 1$ to $t = 1 + h$ as a function of h. Use this graph to estimate the instantaneous velocity at $t = 1$ and compare your answer to the one you obtained in Part 1.

3. Calculate the parachutist's instantaneous velocity at $t = 1$ by using Maple to find the appropriate limit.

4. Plot the graph of $s(t)$ for the first second of the fall and then for the first 20 seconds of the fall. Use this graph to describe the *velocity* of the parachutist as a function of time. Include a discussion of what happens for "large" values of t.

Solved Problem 2.2: Average rates of change (CCH Text 2.2)

Let $f(x) = \dfrac{4x}{4x^2+1}$.

(a) Find the average rate of change of $f(x)$ from $x = 0.25$ to $x = 0.55$.

(b) Find the equation of the corresponding secant line.

(c) Plot the graphs of $f(x)$ and the secant line.

(d) Repeat all of the above steps for $x = 0.25$ to $x = 0.35$. Explain what you observe.

Solution to (a): We first define f in Maple in the usual way:

```
>f := x -> 4*x/(4*x^2 + 1);
```

$$f := x \to 4\,\dfrac{x}{4x^2+1}$$

The average rate of change of f from 0.25 to 0.55 is $\dfrac{f(0.55) - f(0.25)}{0.3}$. At a Maple prompt, therefore, we enter the following to get the result shown.

```
>(f(0.55) - f(0.25))/0.3;
```

$$.651583711$$

Solution to (b): The secant line is the line that passes through the two points $(0.25, f(0.25))$ and $(0.55, f(0.55))$. Its slope is the average rate of change we found in Part (a). Thus the equation of the line is

$$y = 0.651583711(x - 0.25) + f(0.25).$$

To define it in Maple we enter

```
>y := 0.651583711*(x - 0.25) + f(0.25);
```

$$y := .651583711x + .6371040722$$

Solution to (c): Since we are interested in the graphs of y and $f(x)$ near the points in question, we plot these graphs over the interval $[0, 1]$.

```
>plot({y, f(x)}, x = 0..1);
```

The result appears in Figure 2.3a.

Solution to (d): The average rate of change over the interval $[0.25, 0.35]$ is
$$\frac{f(0.35) - f(0.25)}{0.1} = 1.395973156$$
and the secant line is
$$z = 1.395973156(x - 0.25) + f(0.25) = 1.395973156x + .4510067110.$$
(We have called this line z to distinguish it from the previous line found in Part (c).) The graphs of $f(x)$ and the two secant lines may be obtained from Maple as follows.

```
>plot({y, z, f(x)}, x = 0..1);
```

The result is shown in Figure 2.3b.

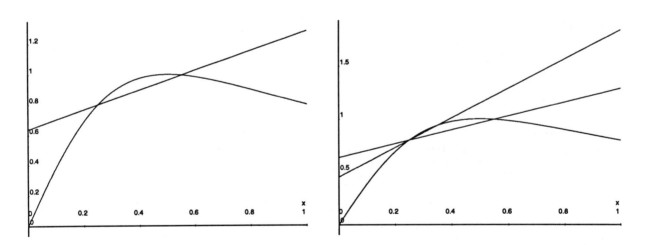

a. b.

Figure 2.3: Plot of $f(x) = \frac{4x}{4x^2+1}$ and the secant lines

The value 0.35 is much closer to 0.25 ($h = 0.1$) than is 0.55 ($h = 0.3$), and we observe that the secant line is now nearly *tangent* to the curve at $(0.25, f(0.25))$. The reader is encouraged to plot secant lines for still smaller values of h (and for negative values as well) to observe the convergence to tangency.

Solved Problem 2.3: Instantaneous rates of change (CCH Text 2.2)

Let $f(x) = \dfrac{4x}{4x^2 + 1}$.

(a) Find the instantaneous rate of change of f at $x = 0.25$ using the definition.

(b) Find the equation of the corresponding tangent line.

(c) Plot the graphs of f and of the tangent line. Zoom in on the point of tangency and explain what you see.

(d) Plot the graph of the difference quotient from Part (a). Explain how you can use this graph to estimate the derivative of f at 0.25.

Solution to (a): We first define f as we did in Solved Problem 2.2. The instantaneous rate of change (or derivative) of f at 0.25 is $\lim\limits_{h \to 0} \dfrac{f(0.25 + h) - f(0.25)}{h}$. To compute this limit, enter

```
>limit((f(0.25 + h) - f(0.25))/h, h = 0);
                    1.920000000
```

Solution to (b): The tangent line is the line through the point $(0.25, f(0.25))$ with slope equal to the instantaneous rate of change that we found in Part (a). So the equation of the tangent line is $y_{\tan} = 1.92(x - 0.25) + f(0.25)$. We define this line in Maple calling it ytan.

```
>ytan := 1.92*(x - 0.25) + f(0.25);
```

$$ytan := 1.92x + .3200000000$$

Solution to (c): Plot both $y_{\tan}$ and f on the same axes with the following Maple command. The result is shown in Figure 2.4a.

```
>plot({ytan, f(x)}, x = -2..2, y = -2..2);
```

You should compare this picture with the graphs of the secant lines plotted in Solved Problem 2.2.

To zoom in on the point of tangency, we just alter the viewing window, giving us Figure 2.4b.

```
>plot({ytan, f(x)}, x = 0.2..0.3);
```

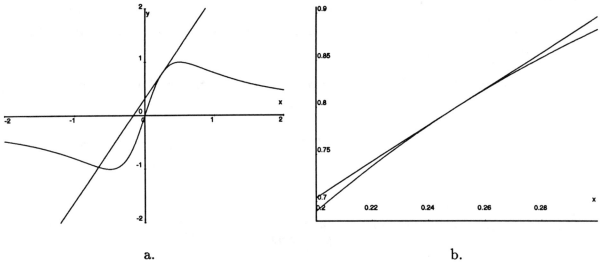

a.   b.

Figure 2.4: Plot of $f(x) = \frac{4x}{4x^2+1}$ and a tangent line

Note that the graph of f and the tangent line seem to be actually merging. This is a demonstration of the principle of *local linearity*. That is, if a function has a derivative at a point, then the function looks and behaves very much like its tangent line if we zoom in close enough. If we continue zooming in by successively making the viewing window smaller, we scarcely will be able to distinguish between the two graphs. (Try plotting the graphs over the domain, 0.23..0.27.)

Solution to (d): Since the derivative of f at 0.25 is $\lim_{h \to 0} \frac{f(0.25+h) - f(0.25)}{h}$, we are interested in what happens to the graph of $\frac{f(0.25+h) - f(0.25)}{h}$ when h is near 0. We first plot a "global" view of this difference quotient. The graph appears in Figure 2.5a.

```
>plot((f(0.25 + h) - f(0.25))/h, h = -3..3);
```

We can estimate the derivative by looking at the second coordinate where the graph of the difference quotient appears to cross the vertical axis. To get a closer view of this point, change the viewing window with the following command. The result is presented in Figure 2.5b.

```
>plot((f(0.25 + h) - f(0.25))/h, h = -0.2..0.2);
```

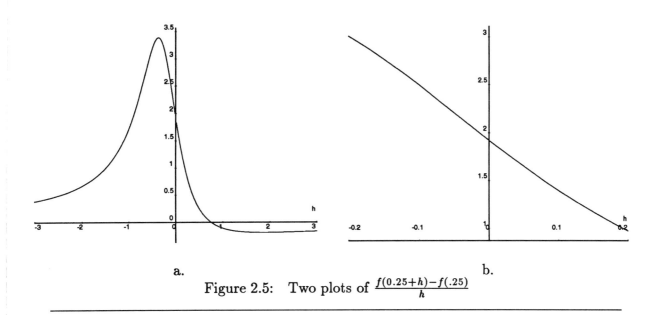

Figure 2.5: Two plots of $\frac{f(0.25+h)-f(.25)}{h}$

Notice that the point where the graph crosses the vertical axis is very close to 2. The actual derivative, as we saw above, is 1.92.

It should be emphasized that the difference quotient $\frac{f(0.25+h)-f(.25)}{h}$ is not defined at $h = 0$, so there is actually a hole in the graph there. What we are estimating is the *limit* of this difference quotient, not its value at $h = 0$.

Remark: It should be pointed out that Maple has a very simple command that will plot the graph and a tangent line together. The command is called **showtangent** and is part of the **student calculus** package. To use it, we must first read this package into Maple's memory. To do this, enter

```
>with(student);
```

When this statement is executed, Maple will print a list of new commands that will be available for your use. Note that one of them is called **showtangent**. Its syntax is

```
>showtangent(f, variable = a, domain, range);
```

where `f` can be either a function or an expression, `a` is the point at which we wish to display the tangent line, and domain and range are optional. In the present example, since we have defined f as a function in Maple, we would enter the command below.

```
>showtangent(f(x), x = 0.25, x = -1..1, y = -2..2);
```

The result is shown in Figure 2.6. Note that, in addition to the tangent line, Maple has also drawn a vertical line from $x = 0.25$ to the point of tangency.

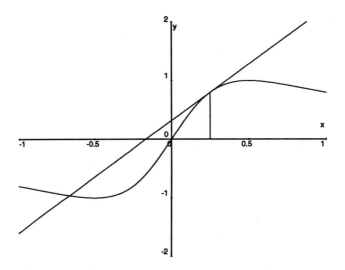

Figure 2.6:   Result of the **showtangent** command

Laboratory Exercise 2.3

Slopes and Average Rates of Change (CCH Text 2.1)

Name _____   Due Date _____

Let $f(x) = 3x - 2x^2$.

1. Find the average rate of change of f from $x = 0.5$ to $x = 0.9$.

2. Find the equation of the corresponding secant line.

3. Plot the graphs of f and the secant line on the same axes.

4. Repeat Parts 1, 2, and 3 for $x = 0.5$ to $x = 0.51$. Explain what you observe.

5. Zoom in on the graph around the point $(0.5, f(0.5))$. Show your plot and explain what you observe about the two graphs in Part 4.

6. Replot the graph of f over the interval $[0, 1]$. Now zoom in on the graph around the point $(0.5, f(0.5))$ until the graph looks like a straight line. Show your plot and explain how you can use this graph to estimate the slope of this "line." (Hint: Move the mouse pointer to the line and click at two different points on it; then observe the first and second coordinates of the points you clicked on.)

Laboratory Exercise 2.4

Tangent Lines and Rates of Change (CCH Text 2.2)

Name _____   Due Date _____

Let $f(x) = 3x - 2x^2$.

1. Find the instantaneous rate of change of f at $x = 0.5$ *using the definition*.

2. Find the equation of the corresponding tangent line.

3. Plot the graphs of f and the tangent line on the same axes. Zoom in on the graph around the point $(0.5, f(0.5))$ until the two graphs are indistinguishable. How close did you have to get?

4. Plot the graph of $\dfrac{f(0.5+h)-f(0.5)}{h}$. Explain how you can use this graph to estimate the derivative of f at 0.5.

Laboratory Exercise 2.5

The Derivative of the Gamma Function (CCH Text 2.2)

Name _____   Due Date _____

Maple knows a function called *the Gamma function*. To see it, just enter GAMMA(x) (all capital letters) at a Maple prompt. Maple will return $\Gamma(x)$. (The symbol Γ is the capital Greek letter *gamma*.) This function is very important in mathematics, and you will encounter it later when you get to the subject of improper integrals. We present it here because we want to study the derivative of a function you have not seen before. Our purpose is to approximate $\Gamma'(1)$ (assuming that $\Gamma(x)$ has a derivative at $x = 1$). Basically, $\Gamma(n)$ is the general form of $(n-1)!$.

1. At a Maple prompt enter each of GAMMA(1), GAMMA(1/2), and GAMMA(3). What do you get?

2. Estimate $\Gamma'(1)$ by evaluating $\dfrac{\Gamma(1+h) - \Gamma(1)}{h}$ with $h = 0.001$.

3. Plot the graph of $\Gamma(x)$ using x = 0..5, y = 0..10. Observe the shape of the graph. Zoom in on the graph near the point $(1, \Gamma(1))$. Show your plot and use it to estimate the slope of this "line." Explain all your steps clearly.

4. Plot the graph of $\dfrac{\Gamma(1+h) - \Gamma(1)}{h}$. Use this graph to estimate the derivative of Γ at $x = 1$. Explain what you did and compare your answer with those you obtained in Parts 2 and 3.

5. Find the equation of the tangent line to $\Gamma(x)$ at $x = 1$ using the estimate for the slope you obtained from Part 4 or from Part 2.

6. Plot the graphs of $\Gamma(x)$ and the tangent line. Zoom in on the point of tangency and explain how these graphs support your estimate of $\Gamma'(1)$.

7. Find the limit of the difference quotient as $h \to 0$ using Maple's `limit` command. What do you get? [1]

---

[1] The actual value of $\Gamma'(1)$ is about -.5772156649. The absolute value of this number is known as *Euler's constant* and is often denoted by γ, the lower case Greek letter gamma. Maple knows this, and you can get it by asking Maple to find the limit as $h \to 0$ of the expression in Part 2.

Solved Problem 2.4: The derivative function (CCH Text 2.3)

Let $f(x) = \dfrac{x}{1+x^2}$.

(a) Use Maple to calculate $f'(x)$ in two ways:
   (i) directly and (ii) by the definition.

(b) On the same axes, plot the graphs of f and f'. Describe the relationship between the sign of f' and the graph of f.

Solution to (a): First, we define f in Maple with

```
>f := x -> x/(1 + x^2);
```
$$f := x \to \frac{x}{1+x^2}$$

There are several ways to find a derivative in Maple. For this problem we will use the `diff` command (short for *differentiate*). See Appendix I for the details on how to use this command.

To get the derivative of f directly, enter

```
>diff(f(x), x);
```
$$\frac{1}{1+x^2} - 2\frac{x^2}{(1+x^2)^2}$$

Now simplify this expression with

```
>simplify(");
```
$$-\frac{-1+x^2}{(1+x^2)^2}$$

This last expression is the derivative of f. To obtain the result using the definition, we need to evaluate the limit as $h \to 0$ of the difference quotient.

```
>limit((f(x + h) - f(x))/h, h = 0);
```
$$-\frac{-1+x^2}{(1+x^2)^2}$$

Note that the two expressions are identical as we expected.

Solution to (b): To plot both f and f' on the same axes, we need to give f' a name in Maple. We can do this with the following command.

```
>df := diff(f(x), x);
```

Maple will return the same result as above, but this time the expression may be referred to by the name df. Now we can plot. The result is shown in Figure 2.7.

```
>plot({f(x), df}, x = -5..5);
```

The graph of f is the one passing through the origin. We observe that the graph of f is increasing where the graph of f' is above the x-axis (on the interval (-1,1)) and that the graph of f is decreasing where the graph of f' is below the x-axis (on the intervals $(-\infty, -1)$ and $(1, \infty)$).

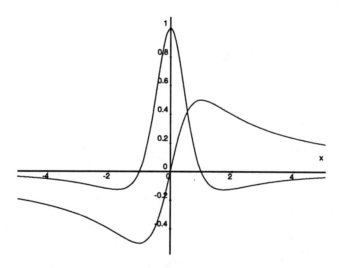

Figure 2.7: $\dfrac{x}{x^2+1}$ and its derivative

<u>Remark:</u> In Appendix II, we have provided a very useful routine for working with difference quotients. It is called DQ. See Appendix II for examples on how to use this function.

Laboratory Exercise 2.6

The Meaning of the Sign of f' (CCH Text 2.3)

Name _____   Due Date _____

This is a graphical demonstration of the following important principle: If f' is positive, then f is increasing. If f' is negative, then f is decreasing. This demonstration requires a color monitor.

1. First define `f := x -> sin(x)` in Maple. Next, at a Maple prompt, enter the following lines <u>exactly</u> as they appear here. Be sure to pay attention to the syntax, upper and lower case letters, and to press return after each line.

```
>p := proc(x) if D(f)(x) > 0 then f(x) else NOP fi; end;
>q := proc(x) if D(f)(x) < 0 then f(x) else NOP fi; end;
```

Once you have done this, enter the following and explain what you see.

```
>plot({p, q}, 0..Pi);
```

2. In Maple, `D(f)` is the notation for the derivative function, f', and `D(f)(x)` is just $f'(x)$. Explain what the above formula is doing and how it works.

3. Repeat the above experiment for $x^3 - x$.

Laboratory Exercise 2.7

Recovering f from f' (CCH Text 2.5)

Name _____   Due Date _____

Suppose f is a function with the property that $f(0) = 0$ and $f'(x) = \ln(x^4 - x^2 + 0.5)$.

1. Plot the graph of f' and determine the intervals where it is positive and where it is negative.

2. Use Maple to calculate f''.

3. Plot the graph of f'' and determine the intervals where it is positive and where it is negative.

4. Use the information in Parts 1 and 3 to sketch the graph of f. Your graph should accurately reflect this information.

Solved Problem 2.5: Calculating limits graphically (CCH Text 2.7)

Use graphs to estimate the value of the two limits $\lim_{x \to 0}(1+3x)^{\frac{1}{x}}$ and $\lim_{x \to 0} \sin(\frac{1}{x})$. Use Maple to calculate the limits and compare them with your estimates.

Solution: First, we define the function $f(x) = (1+3x)^{\frac{1}{x}}$ in Maple and then plot it. See Figure 2.8a.

```
>f := x -> (1 + 3*x)^(1/x);
```

$$f := x \to (1+3x)^{(\frac{1}{x})}$$

```
>plot(f(x), x= -0.1..0.1);
```

(Why did we choose the viewing window as we did?)

If we move the mouse pointer to the point where the curve crosses the vertical axis and click, we can get a reasonable estimate for the limit. This estimate could be improved by zooming in on the graph near this point.

To compute the limit exactly, we use Maple's `limit` function.

```
>limit(f(x), x = 0);
```

$$e^3$$

Is this result a surprise? Perhaps you can relate the given limit to the more familiar $\lim_{x \to \infty}(1+\frac{3}{x})^x$. Finally, we use `evalf` to obtain the decimal approximation of e^3.

```
>evalf(");
```

$$20.08553692$$

We note that this result is very close to the one we obtained graphically above.

For the second part of this problem, we again begin by defining the function in Maple, and then plot over the domain $[-\pi, \pi]$. The result appears in Figure 2.8b. Note that when x is near 0, the graph appears to oscillate wildly and does not get close to any single value. If you zoom in near the origin, the oscillation will become even more apparent. We conclude that the limit does not exist. However, we can attempt to compute this limit using Maple:

```
>limit(f(x) , x = 0);
```
$$-1 \mathrel{..} 1$$

This is Maple's way of trying to tell us that the limit is all values from -1 to 1. Since this is absurd, the limit does not exist. *Although our assertion that the limit does not exist is correct, this is not a verification; it only means that Maple cannot find an answer.*

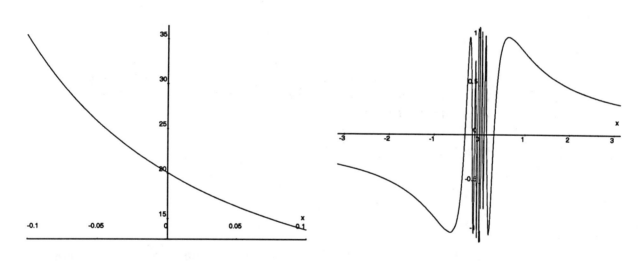

a.   b.
Figure 2.8:   Plot of $f(x) = (1+3x)^{\frac{1}{x}}$ and $f(x) = \sin(\frac{1}{x})$

Laboratory Exercise 2.8

Using Graphs to Estimate Limits (CCH Text 2.7)

Name _____   Due Date _____

Estimate the value of each of the following limits by looking at a graph; then verify your estimate where possible by asking Maple to calculate the limit.

1. $\lim\limits_{x \to 0} \dfrac{3^x - 1}{x}$

2. $\lim\limits_{x \to \frac{\pi}{2}} (x - \dfrac{\pi}{2}) \tan x$

3. $\lim\limits_{x \to 0} \dfrac{2^x - 1}{|x|}$

Chapter 3
Key Concept: The Definite Integral

The definite integral is defined as a limit of Riemann sums. Throughout this book all Riemann sums are assumed to use subintervals of equal length.

Solved Problem 3.1: Measuring distance (CCH Text 3.1)

The downward velocity of a parachutist t seconds after jumping from an airplane is given by $v(t) = 20(1 - e^{-1.6t})$ feet per second. Suppose the parachutist jumps from an airplane and lands on the ground 45 seconds later.

(a) Approximate the total distance the parachutist has fallen using a left-hand Riemann sum with 50 subintervals.

(b) Approximate the total distance the parachutist has fallen using a right-hand Riemann sum with 50 subintervals.

(c) What is the maximum error in your approximation if you use either of the two estimates above?

(d) What is the maximum error if you use the average of the approximations calculated in Parts (a) and (b) above?

Solution to (a): First define the velocity function v in Maple with

```
>v := t -> 20*(1 - exp(-1.6*t));
```
$$v := t \to 20 - 20e^{(-1.6t)}$$

The left-hand sum for $v(t)$ on $[0, 45]$ with 50 subintervals is $\sum_{i=0}^{49} v(t_i)\Delta t$, where $t_i = \frac{45i}{50}$ and $\Delta t = \frac{45}{50}$. Maple's **sum** command allows us to compute this sum quickly.

```
>sum(v(45*i/50)*(45/50), i = 0..49);
```
$$876.4111456$$

Solution to (b): The right-hand sum is $\sum_{i=1}^{50} v(t_i)\Delta t$. To get this sum, we re-execute the above command by changing the appropriate values for i.

```
>sum(v(45*i/50)*(45/50), i = 1..50);
                                894.4111456
```

Solution to (c): In Figure 3.1, we have plotted the graph of $v(t)$ with

```
>plot(v(t), t = 0..4);
```

Since this graph is increasing, the left-hand sum is an underestimate of the total distance traveled, and the right-hand sum is an overestimate. The true distance traveled will lie somewhere between the left-hand and right-hand sums calculated in Parts (a) and (b) above. Whether we choose 876.4111457 or 894.4111456 as an answer, the error can be no more than their difference, which is approximately 18 feet.

Solution to (d): The average of the left and right sums, 885.4111456, lies halfway between the two. Since the true distance also lies between the left and right sums, the error can be no more than half their difference, about 9 feet, when we use the estimate 885.4111456 feet.

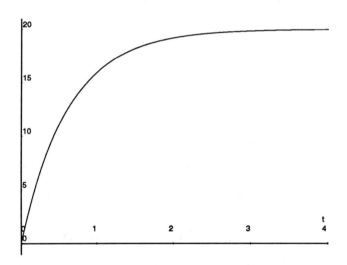

Figure 3.1: Velocity for a falling parachutist

Laboratory Exercise 3.1

An Asteroid (CCH Text 3.1)

Name _____ Due Date _____

An asteroid is falling toward earth so that its velocity t hours after it was first observed at time $t = 0$ is given by

$$v(t) = \frac{1830000}{(8760 - t)^{\frac{1}{3}}} \text{ kilometers per hour.}$$

Determine how far the asteroid travels during the first 6 months after it was first observed. (Take care that your units are correct.)

1. Plot the graph of $v(t)$. (You should alter the viewing window to produce a meaningful graph. To do this, think about how many hours there are in a month.)

2. Use a left-hand sum with 1000 subintervals to approximate the distance traveled by the asteroid during the first 6 months. (We suggest that you nest the `sum` command within an `evalf` command; that is, in `evalf( ... )`, replace the three dots with the sum command.)

3. Use a right-hand sum with 1000 subintervals to approximate the distance traveled by the asteroid during the first 6 months.

4. Using your work in Parts 2 and 3, make the best estimate you can of the distance the asteroid traveled in the first 6 months. Include an upper bound on the error.

5. When will the asteroid strike the earth? (<u>Hint:</u> Consider the graph in Part 1 and the definition of $v(t)$. They display interesting behavior near a certain point. Discuss what you observe.) Also discuss the validity of the formula for $v(t)$ when the asteroid is very near the earth.

Laboratory Exercise 3.2

A Falling Water Table (CCH Text 3.1)

Name _____   Due Date _____

t days after a heavy rain, the water table at a point near a drainage ditch falls at a rate of
$$v(t) = \frac{10e^{-\frac{1}{t+0.1}}}{(t+0.1)^{\frac{3}{2}}} \text{ inches per day.}$$

1. Use a left-hand Riemann sum with 100 subintervals to estimate how far the water table falls in three days.

2. Plot $v(t)$. Can you determine if the number you obtained in Part 1 is an overestimate or an underestimate of the distance the water table has fallen? Explain.

3. Use the graph to determine approximately where $v(t)$ is increasing and where it is decreasing. Explain how this information can be used to produce Riemann sums that give an overestimate and an underestimate of the true distance the water table has fallen.

Solved Problem 3.2: Calculating Riemann sums (CCH Text 3.2)

Let $f(x) = x^2$.

(a) Calculate the left-hand Riemann sum for $f(x)$ on $[0, 2]$ using 50 subintervals.

(b) Calculate the right-hand Riemann sum for $f(x)$ on $[0, 2]$ using 50 subintervals.

(c) In order to approximate $\int_0^2 x^2\, dx$ to one decimal place by a left-hand sum, how many subintervals must be used? What is the left-hand sum for that number of subintervals?

(d) In order to approximate $\int_0^2 x^2\, dx$ to one decimal place by the average of the left-hand and right-hand sums, how many subintervals must be used? What is this average for that number of subintervals?

(e) Use Maple's student calculus package to plot the graph of $f(x) = x^2$ along with (i) a left-hand sum of 10 subintervals and (ii) a right-hand sum of 10 subintervals.

Solution to (a): The procedure for entering a Riemann sum is the same as that described in Solved Problem 3.1. The sum we want is $\sum_{i=0}^{n-1} f(x_i)\Delta x$, where $f(x) = x^2$, $x_i = \dfrac{2i}{n}$, $\Delta x = \dfrac{2}{n}$, and $n = 50$. We first define f in Maple with

```
>f := x -> x^2;
```
$$f := x \to x^2$$

and then use the sum command as in Solved Problem 3.1 together with the evalf command.

```
>evalf(sum(f(2*i/50)*(2/50), i = 0..49));
                    2.587200000
```

Solution to (b): The right-hand sum is $\sum_{i=1}^{50} f\left(\dfrac{2i}{50}\right)\dfrac{2}{50}$. To obtain this sum, we execute the command in Part (a) again with the range for i changed to 1..50.

```
>evalf(sum(f(2*i/50)*(2/50), i = 1..50));
                    2.747200000
```

Solution to (c): The key idea is that for a monotone function, the difference between the left-hand and right-hand sums is $|f(b) - f(a)|\Delta x$. Therefore, since the integral is between the two sums, this places a bound on the error between the left-hand sum (or right-hand sum) and the integral. If n is the number of subintervals, then $\Delta x = \dfrac{2}{n}$, so $|f(2) - f(0)|\Delta x = 4\dfrac{2}{n} = \dfrac{8}{n}$. For one decimal place accuracy, we must have $\dfrac{8}{n} < 0.05$. Thus $n > \dfrac{8}{0.05} = 160$. If we use 161 subintervals we can guarantee one-place accuracy.

We need to calculate $\sum\limits_{n=0}^{160} f\left(\dfrac{2i}{161}\right)\dfrac{2}{161}$. This can be done in much the same way we computed the sums in Parts (a) and (b).

```
>evalf(sum(f(2*i/161)*(2/161), i = 0..160));
                     2.641873385
```

Solution to (d): Since the integral is between the two sums and their difference is $|f(b) - f(a)|\Delta x$, then their average is within $\dfrac{1}{2}|f(b) - f(a)|\Delta x$ of the integral. Therefore, we want n to be large enough to make $\dfrac{1}{2}|f(b) - f(a)|\Delta x < 0.05$. That is, we need a value for n so that $\dfrac{1}{2}4\dfrac{2}{n} < 0.05$, or $n > 80$. We can guarantee one-place accuracy if we use 81 subintervals. The left-hand and right-hand sums appear below together with their average.

```
>a := evalf(sum(f(2*i/81)*(2/81), i = 0..80));
                     a := 2.617487172
```

```
>b := evalf(sum(f(2*i/81)*(2/81), i = 1..81));
                     b := 2.716252604
```

```
>(a + b)/2;
                     2.666869888
```

Solution to (e): To use Maple's `student calculus` package, we must first read it into Maple's memory. Do this using the following command:

```
>with(student);
```

When this command is executed, Maple will display a list of additional commands that are now available to us. The two we will be interested in are `leftbox` and `rightbox`. Other commands in this package will be used in Solved Problem 3.3.

To get a picture of the left-hand sum with 10 subintervals, we use `leftbox` as follows. (Pay close attention to the syntax.)

>leftbox(f(x), x = 0..2, 10);

Getting the picture of the right-hand sum with 10 subintervals is just as easy if we use `rightbox` instead of `leftbox` and re-execute the above command. The results of these commands are displayed in Figure 3.2a and 3.2b.

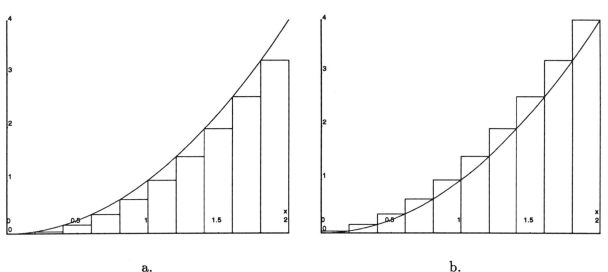

a. b.

Figure 3.2: Plot of the left-hand and right-hand Riemann sums

Solved Problem 3.3: Limits of Riemann sums (CCH Text 3.2)

Let $f(x) = x^2$.

(a) Calculate the left-hand sum for $f(x)$ on $[1,2]$ using n subintervals. Calculate the limit of the left-hand sum as $n \to \infty$, and explain the meaning of the limit you have calculated.

(b) Calculate the right-hand sum for $f(x)$ on $[1,2]$ using n subintervals. Calculate the limit of the right-hand sum as $n \to \infty$, and explain the meaning of the limit you have calculated.

Solution to (a): Define the function f in Maple with

```
>f := x -> x^2;
```
$$f := x \to x^2$$

Now we read the **student calculus** package into Maple's memory.

```
>with(student);
```

This time, we will be interested in the **leftsum**, **rightsum**, and **value** procedures provided in this package. The left-hand sum for $f(x)$ on $[1,2]$ with n subintervals is $\sum_{i=0}^{n-1} f(x_i)\Delta x$, where $x_i = 1 + \dfrac{i}{n}$ and $\Delta x = \dfrac{1}{n}$. We can get this sum using the **leftsum** command as follows:

```
>leftsum(f(x), x = 1..2, n);
```
$$\frac{\sum_{i=0}^{n-1}(1+\frac{i}{n})^2}{n}$$

Notice that Maple has returned the value to us symbolically. To get the value of this sum, use the **value** command on it.

```
>value(");
```
$$\frac{\frac{7}{3}n - \frac{3}{2} + \frac{1}{6}\frac{1}{n}}{n}$$

We now can find the limit of this last expression as $n \to \infty$ by using Maple's `limit` command.

```
>limit(", n = infinity);
```
$$\frac{7}{3}$$

According to the definition of the integral, this is the exact value of $\int_1^2 x^2\, dx$.

Solution to (b): The right-hand sum for $f(x)$ on $[1,2]$ is $\sum_{i=1}^{n} f(x_i)\Delta x$, where $x_i = 1 + \dfrac{i}{n}$ and $\Delta x = \dfrac{1}{n}$. We can get this sum by using the `rightsum` command as follows:

```
>rightsum(f(x), x = 1..2, n);
```
$$\frac{\sum_{i=1}^{n}(1+\frac{i}{n})^2}{n}$$

Now use `value` to get the value of this sum as we did in Part (a). Note that the expression that Maple returns is somewhat more complex than it was with `leftsum`. We can simplify it further using `simplify`.

```
>simplify(");
```
$$\frac{1}{6}\frac{14n^2 + 9n + 1}{n^2}$$

Lastly, we use `limit` to compute the limit of the above expression as $n \to \infty$:

```
>limit(", n = infinity);
```
$$\frac{7}{3}$$

The limit here and the one in Part (a) agree even though the left-hand and right-hand sums are different. As mentioned in Part (a), this limit is the exact value of $\int_1^2 x^2\, dx$.

Laboratory Exercise 3.3

Calculating Riemann Sums (CCH Text 3.2)

Name _____   Due Date _____

Let $f(x) = \frac{1}{4}(6 + 7x - x^3)$.

1. Calculate the left-hand Riemann sum for $f(x)$ on $[-1, 1]$ using 30 subintervals.

2. Calculate the right-hand Riemann sum for $f(x)$ on $[-1, 1]$ using 30 subintervals.

3. In order to approximate $\int_{-1}^{1} f(x)\, dx$ to one decimal place by a right-hand sum, how many subintervals must be used? What is the right-hand sum for that number of subintervals?

4. In order to approximate the integral $\int_{-1}^{1} f(x)\,dx$ to one decimal place by the average of the left-hand and right-hand sums, how many subintervals must be used? What is this average for that number of subintervals?

5. Use the student calculus package to plot the graph of $f(x)$ on the interval $[-1, 3]$ with a left-hand sum of 10 subintervals. In Solved Problem 3.2 the left-hand sum stayed underneath the graph. Did that happen here? Explain.

6. Use the student calculus package to plot the graph of $f(x)$ on the interval $[-1, 3]$ with a right-hand sum of 10 subintervals. In Solved Problem 3.2 the right-hand sum stayed above the graph. Did that happen here? Explain.

Laboratory Exercise 3.4

Limits of Riemann Sums (CCH Text 3.2)

Name _____   Due Date _____

Let $f(x) = \frac{1}{4}(6 + 7x - x^3)$.

1. Calculate the left-hand sum for $f(x)$ on $[2, 3]$ using n subintervals.

2. Calculate the right-hand sum for $f(x)$ on $[2, 3]$ using n subintervals.

3. Calculate the limits of the left-hand and right-hand sums as $n \to \infty$ and explain the meanings of the limits you have calculated.

Laboratory Exercise 3.5

Estimating Integrals with Riemann Sums (CCH Text 3.2)

Name _____   Due Date _____

In each of the following, (a) find the number of subintervals necessary for the average of the left and right sums to approximate the integral to one decimal place, and (b) find the average of the left and right sums for that number of subintervals.

1. $\int_1^2 \sin(\ln t)\, dt$

2. $\int_{\pi/2}^{\pi} \sqrt{\sin t}\, dt$

Laboratory Exercise 3.6

Riemann Sums and the Fundamental Theorem (CCH Text 3.4)

Name _____   Due Date _____

1. Calculate the left-hand sum for $f(t) = t^2$ on $[0, x]$ using n subintervals.

2. Calculate the limit of the left-hand sum above as $n \to \infty$.

3. Calculate $\int_0^x t^2 \, dt$ using the Fundamental Theorem of Calculus.

4. What do you observe about your answers in Parts 2 and 3?

5. Repeat Parts 1 through 4 above for $f(t) = e^t$.

Calculating Definite Integrals with Maple (CCH Text 3.4)

Maple can evaluate many definite integrals exactly. We will illustrate the rather simple procedure for doing so with $\int_0^\pi \sin x \, dx$. Enter the following at a Maple prompt:

```
>int(sin(x), x = 0..Pi);
```
$$2$$

Note first that this answer is the one we would have obtained had we used the Fundamental Theorem of Calculus. Since $\dfrac{d}{dx}(-\cos x) = \sin x$, $-\cos x$ is an antiderivative of $\sin x$. Applying the Fundamental Theorem of Calculus gives

$$\int_0^\pi \sin x \, dx = (-\cos \pi) - (-\cos 0) = 2.$$

Notice also that we did not have to define sine as a function first, although we could have. For example, we could have proceeded by defining a function first, then executing the above `int` command with $f(x)$ in place of $\sin(x)$.

```
>f := x -> sin(x);
```
$$f := \sin$$

```
>int(f(x), x = 0..Pi);
```
$$2$$

Maple can calculate many definite integrals, and in some cases you may not understand the result. For example, let's look at $\int_0^\pi \sin(x^2) \, dx$.

```
>int(sin(x^2), x = 0..Pi);
```
$$\frac{1}{2}\sqrt{2}\sqrt{\pi} \; \text{FresnelS}\left(\sqrt{2}\sqrt{\pi}\right)$$

(FresnelS is called the Fresnel sine integral.) When Maple returns our request for the value of the integral in terms of a function that is unknown to us, such as in this example, one recourse is to ask Maple to provide a numerical value. We do this now with

```
>evalf(");
```
$$.7726517130$$

Thus, to 10 decimal places, $\int_0^\pi \sin(x^2)\,dx = 0.7726517130$.

In some cases, Maple will not know how to evaluate an integral. For example, Maple cannot find $\int_0^\pi \sin(x^3)\,dx$. If you ask Maple to do so, it will "think" for a while, then just return the integral to you.

>int(sin(x^3), x = 0..Pi);

$$\int_0^\pi \sin(x^3)\,dx$$

Yet Maple can still produce an approximation:

>evalf(");

.4158338147

For most of the definite integrals we will encounter, Maple will return a value or familiar expression. If it does not, try asking for a numerical answer using **evalf**.

Laboratory Exercise 3.7

Calculating Areas (CCH Text 3.3)

Name _____   Due Date _____

Produce pictures of the following regions and calculate their areas. Express your answers as integrals and ask Maple to evaluate them. Where exact answers cannot be found, provide approximations.

1. The region between $\sin x$ and the x-axis from $x = 0$ to $x = 2\pi$. (Be careful. The correct answer is *not* $\int_0^{2\pi} \sin x \, dx$. Remember the integral gives the area over intervals where the function is positive, but gives the negative of the area over intervals where the function is negative.)

2. The region inside the unit circle $x^2 + y^2 = 1$ and above the graph of $y = x^2$.

3. The region inside the unit circle and above the graph of $y = x^3$. (<u>Hint:</u> If you examine the picture carefully, you can find the exact answer without calculus or a computer.)

Laboratory Exercise 3.8

The Average Value of a Function and the Fundamental Theorem
(CCH Text 3.4)

Name _____   Due Date _____

1. Find the average value of x^4 on $[a, b]$.

2. Find the limit of your answer in Part 1 as $b \to a$.

3. Use Maple to try to find the average value of $\sin(x^4)$ on the interval $[a, b]$. What does Maple's response tell you about its ability to find this integral?

4. Use Maple's `limit` command to find the limit of the integral in Part 3 as $b \to a$.

5. In Part 2 it was not hard to see how Maple could find the limit because it could find an explicit formula for the integral. However, even though it could not evaluate the integral in Part 3, it was still able to find the limit of the average values in Part 4. Pick a variable you have not as yet used, suppose it is g, so that $g(x)$ denotes an arbitrary function in Maple, then ask Maple to evaluate $\lim_{b \to a} \dfrac{\int_a^b g(x)\,dx}{b-a}$. What did you get?

6. By using graphs and the concept of the "average value of a function," explain how Maple was able to get the answer to Part 5.

Chapter 4
Short Cuts to Differentiation

Maple can find the derivative of virtually any combination of the usual elementary functions. Its answers occasionally may look different from the ones you get by hand, but that doesn't mean you are wrong. Just try simplifying a bit more.

Solved Problem 4.1: Difference quotients and derivatives (CCH Text 4.2)

(a) Use Maple to verify that $\frac{d}{dx}x^{10} = 10x^9$.

(b) Use Maple to **simplify** the difference quotient $\frac{(x+h)^{10} - x^{10}}{h}$ and explain how the simplified form can be used to verify that $\frac{d}{dx}x^{10} = 10x^9$.

Solution to (a): The point of this part of the exercise is just to show the steps involved in using Maple to calculate derivatives. The simplest way to do this is to use Maple's **diff** (for *differentiate*) command as follows.

>diff(x^10, x);

$$10x^9$$

Solution to (b): We can enter and simplify the difference quotient in one step with

>simplify(((x + h)^10 - x^10)/h);

$$10x^9 + 45x^8h + 120x^7h^2 + 210x^6h^3 + 252x^5h^4 + 210x^4h^5 + 120x^3h^6 + 45x^2h^7 + 10xh^8 + h^9$$

Notice that each term of the result, except the first, contains a factor of h. Thus, in the limit, each term except $10x^9$ vanishes.

>limit(", h = 0);

$$10x^9$$

Thus the result here and that of Part (a) are identical.

Note also that we did not need to simplify the difference quotient $\dfrac{(x+h)^{10} - x^{10}}{h}$ in order for Maple to find its limit as the following two commands show.

>((x+h)^10-x^10)/h;
$$\frac{(x+h)^{10} - x^{10}}{h}$$

>limit(", h = 0);
$$10x^9$$

However, this is not true in all the difference quotients we will come across. If Maple cannot find the limit of a difference quotient directly (assuming it exists), try simplifying the result, then take the limit.

Remark: In Appendix II, we have provided a very useful routine for working with difference quotients. It is called DQ. See Appendix II for examples on how to use this function.

Laboratory Exercise 4.1

Difference Quotients and the Derivative (CCH Text 4.2)

Name _____   Due Date _____

For each of the following functions, (a) use Maple to calculate the derivative directly, and (b) ask Maple to simplify the appropriate difference quotient and explain how it can be used to verify Maple's answer from Part (a).

1. $f(x) = \dfrac{1}{x^5}$

2. $f(x) = \dfrac{x}{x+1}$

3. $f(x) = \dfrac{x^2}{x+5}$

Solved Problem 4.2: The product rule (CCH Text 4.4)

Verify the product rule and the Fundamental Theorem of Calculus with Maple.

Solution: The point of this exercise is not really to prove that Maple knows the product rule and the Fundamental Theorem of Calculus, but to show you how to use Maple to work with derivatives of arbitrary functions.

First, we may define the product of two arbitrary functions f and g as follows:

>y := f(x)*g(x);
$$y := f(x)g(x)$$

Now we can differentiate the product y with respect to x using Maple's `diff` command.

>diff(y, x);
$$\left(\frac{\partial}{\partial x}f(x)\right)g(x) + f(x)\left(\frac{\partial}{\partial x}g(x)\right)$$

If you are using Release 1 or the DOS version, Maple will use the standard derivative notation, $\frac{d}{dx}$, that you may be used to seeing. For Release 2 under Windows© or on other computers, Maple uses another notation that is usually used to denote derivatives of functions of more than one variable. This notation, $\frac{\partial}{\partial x}$ is called the *partial* derivative notation. Maple will use it consistently to indicate derivatives so when you see it on your screen, you will know what it means – the derivative of $f(x)$. With this in mind, Maple has given us the product rule.

To illustrate the Fundamental Theorem of Calculus, we will integrate the derivative of an arbitrary function $F(x)$. First we give its derivative a name:

>dF := diff(F(x), x);
$$dF := \frac{\partial}{\partial x}F(x)$$

Now we integrate dF from $x = a$ to $x = b$.

>int(dF, x = a..b);
$$F(b) - F(a)$$

Thus Maple does indeed know the Fundamental Theorem of Calculus.

Laboratory Exercise 4.2

The Quotient Rule and the Chain Rule (CCH Text 4.5)

Name _____   Due Date _____

Consider f and g to be arbitrary functions in the following exercises.

1. Ask Maple to verify the quotient rule by differentiating $f(x)/g(x)$. Is Maple's answer the same as that in your textbook? If not, show that the two are the same.

2. Find the derivative of $f^2(x)$ by hand and then ask Maple to find its derivative. Are your answers the same?

3. Find the derivative of $f(x^2)$ by hand and then ask Maple to find its derivative. Are your answers the same? Explain what you observe.

4. Find the derivative of $\sin(f(x))$ by hand and then ask Maple to find its derivative. Are your answers the same?

5. Find the derivative of $f(\sin(x))$ by hand and then ask Maple to find its derivative. Are your answers the same? Explain what you see.

6. Based on your observations in the last four parts, discuss *in general* what Maple seems to know about the chain rule and what it does not seem to know.

7. What will you expect to get if you ask Maple to find the derivative of $f(g(x))$?

8. Ask Maple to find the derivative of $\int_a^x f(t)\,dt$. Prove that Maple's answer is correct. (<u>Hint:</u> Let $f(t) = g'(t)$.)

Solved Problem 4.3: Implicit differentiation (CCH Text 4.8)

Suppose $\sin(x+y) + \sqrt{x^2+y^2} = xy + 4$.

(a) Find $\dfrac{dy}{dx}$ by implicit differentiation.

(b) Find $\dfrac{dy}{dx}$ when $x = \sqrt{2}$ and $y = -\sqrt{2}$.

(c) Find the equation of the tangent line to the graph of the given equation at $(\sqrt{2}, -\sqrt{2})$. Use it to approximate the value of y when $x = 1.5$.

Solution to (a): The first step is tell Maple that y is a function of x. If we do not do this, Maple will treat y as a constant when we differentiate. We do this with

>y := f(x);
$$y := f(x)$$

Now enter the given equation into Maple's memory:

>eq := sin(x + y) + sqrt(x^2 + y^2) = x*y + 4;
$$eq := \sin(x + f(x)) + \sqrt{x^2 + f(x)^2} = xf(x) + 4$$

Notice that Maple has replaced each occurrence of y with $f(x)$. The next step is to differentiate this last equation with respect to x using the `diff` command.

>deq := diff(eq, x);
$$deq := \cos(x + f(x))\left(1 + \left(\frac{\partial}{\partial x}f(x)\right)\right) + \frac{1}{2}\frac{2x + 2f(x)\left(\frac{\partial}{\partial x}f(x)\right)}{\sqrt{x^2 + f(x)^2}} = f(x) + x\left(\frac{\partial}{\partial x}f(x)\right)$$

Keep in mind that ultimately we wish to determine $\frac{d}{dx}f(x)$. To do this we need to replace each occurrence of $\frac{d}{dx}f(x)$ in this last equation with a variable for which Maple can solve. It can be any variable not already used. To this end, we define $z = \frac{d}{dx}f(x)$ in Maple with

>z := diff(f(x), x);
$$z := \frac{\partial}{\partial x}f(x)$$

Note that this last command makes z synonymous with $\frac{d}{dx}f(x)$ so that when we substitute for z in the equation we will actually be substituting for $\frac{d}{dx}f(x)$. We do this with

```
>subs(z = df, deq);
```

$$\cos(x+f(x))(1+df) + \frac{1}{2}\frac{2x+2f(x)df}{\sqrt{x^2+f(x)^2}} = f(x) + xdf$$

The variable we now wish to solve for is *df*. Do this with

```
>solve(", df);
```

$$-\frac{\cos(x+f(x)) + \dfrac{x}{\sqrt{x^2+f(x)^2}} - f(x)}{\cos(x+f(x)) + \dfrac{f(x)}{\sqrt{x^2+f(x)^2}} - x}$$

Simplifying will make this complex fraction look somewhat neater.

```
>dy := simplify(");
```

$$dy := \frac{\cos(x+f(x))\sqrt{x^2+f(x)^2} + x - f(x)\sqrt{x^2+f(x)^2}}{-\cos(x+f(x))\sqrt{x^2+f(x)^2} - f(x) + x\sqrt{x^2+f(x)^2}}$$

Solution to (b): Observe that the given values $x = \sqrt{2}$ and $y = -\sqrt{2}$ actually satisfy the given equation which you can check by hand. We must replace $f(x)$ by $-\sqrt{2}$ and x by $\sqrt{2}$. *It is important that the substitutions be carried out in this order.*

```
>subs(f(x) = -sqrt(2), x = sqrt(2), dy);
```

$$\frac{\cos(0)\sqrt{4} + \sqrt{2} + \sqrt{2}\sqrt{4}}{-\cos(0)\sqrt{4} + \sqrt{2} + \sqrt{2}\sqrt{4}}$$

Now simplify once more to get the answer for this part.

```
>m := simplify(");
```

$$m := \frac{2+3\sqrt{2}}{-2+3\sqrt{2}}$$

Solution to (c): The tangent line passes through the point $(\sqrt{2}, -\sqrt{2})$ and has slope $\dfrac{2+3\sqrt{2}}{-2+3\sqrt{2}}$. Thus its equation is $y = \dfrac{2+3\sqrt{2}}{-2+3\sqrt{2}}(x - \sqrt{2}) - \sqrt{2}$. The value $x = 1.5$ is close to $\sqrt{2}$ so the tangent line should yield a good approximation to the true functional value. We define this tangent line in Maple, and then substitute $x = 1.5$ with the following two commands:

```
>ytan := m*(x - sqrt(2)) - sqrt(2);
```

$$ytan := \frac{(2+3\sqrt{2})(x-\sqrt{2})}{-2+3\sqrt{2}} - \sqrt{2}$$

```
>evalf(subs(x = 1.5, ytan));
```
$$-1.175417436$$

Thus the approximate value for y when $x = 1.5$ is -1.175417436.

Remark: You are encouraged to plot a graph of the given equation and the tangent line. Use `implicitplot` for this purpose. Recall that we used this command in Solved Problem 1.4.

Laboratory Exercise 4.3

Implicit Differentiation (CCH Text 4.8)

Name _____   Due Date _____

Assume y is a function of x such that $ye^{xy^2} + ye^x + xe^y = 2$.

1. Find $\dfrac{dy}{dx}$ by implicit differentiation.

2. Verify that $x = 0$ and $y = 1$ satisfy the given equation.

3. Find $\dfrac{dy}{dx}$ when $x = 0$ and $y = 1$.

4. Find the equation of the tangent line at $(0,1)$ and use it to approximate the value of $y(0.15)$.

5. Check the accuracy of your approximation in Part 4 by substituting $x = 0.15$ into the original equation and solving for y.

6. Plot the given equation and the tangent line. Use `implicitplot` for this purpose.

Chapter 5
Using the Derivative

Since the derivative measures the rate of change of a function, it can help solve many different problems, especially those that involve determining where a function reaches a maximum or minimum.

A word of advice concerning word problems: When a problem involves physical units, be sure your answer does too. Also, always ask yourself whether your answer makes sense. If you are supposed to find the length of a pencil, and you get an answer such as 200 miles, -0.22 meter, or $3 + i\sqrt{5}$ feet, you should suspect something is wrong.

Solved Problem 5.1: Maxima and minima (CCH Text 5.1)

Let $f(x) = e^{2x} - 3e^x$.

(a) Use the graph of f to estimate all local maxima and minima of f.

(b) Plot the graph of f' and explain how the graph supports your conclusions in Part (a).

(c) Use the derivative of f to find the *exact* values of the maxima and minima.

Solution to (a): First, we define f in Maple:

```
>f := x -> exp(2*x) - 3*exp(x);
```

$$f := x \to e^{(2x)} - 3e^x$$

Then we plot f with the following command. The result appears in Figure 5.1a.

```
>plot(f(x), x = -2..2, y = -3..3);
```

We see that the graph has a single local minimum, which also *appears* to be a global minimum, and there *seems* to be no local maximum. Notice that we are careful not to make the last two assertions definitively because we aren't sure yet whether something to the contrary may be happening off the computer screen. Part (c) will resolve this. Now move the mouse pointer to the apparent minimum and click once to view the approximate coordinates of the point.

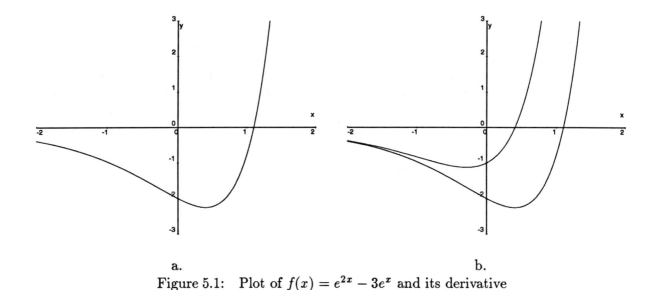

a.   b.

Figure 5.1: Plot of $f(x) = e^{2x} - 3e^x$ and its derivative

Solution to (b): We find the derivative of f and give it the name *df* using Maple's `diff` command:

```
>df := diff(f(x), x);
```
$$df := 2e^{(2x)} - 3e^x$$

Now we plot *df* along with f on the same coordinate axes. The result of the following command appears in Figure 5.1b.

```
>plot({df, f(x)}, x = -2..2, y = -3..3);
```

We observe that the graph of the derivative crosses the x-axis, changing from positive to negative at about $x = 0.4$. This shows that f decreases up to this point, reaches a local minimum, and then starts to increase. If you move the mouse pointer to this crossing point and click, you may get a more accurate estimate of the x coordinate of the minimum than you obtained in Part (a).

Solution to (c): To get the exact value of the minimum, we must find the critical values of f, namely, the zeros of its first derivative. To do so, we ask Maple to solve the equation $df = 0$ for x.

`>solve(df = 0, x);`

$$\ln(\frac{3}{2})$$

Because this is the *only* critical point, we can now be sure that the graph indeed has a global minimum at this point and that there is no local maximum.

To find the y value, ask Maple to find the value of f at this point.

`>f(ln(3/2));`

$$\frac{-9}{4}$$

(You should be able to perform this computation by hand. You should also be able to solve the equation $df = 0$ by hand. Try it.) You can use Maple's `evalf` command to get decimal values for $\ln(3/2)$ and for $-9/4$ to compare them with the x and y estimates of the coordinates of the minimum obtained in Part (a).

Solved Problem 5.2: Critical points and extrema (CCH Text 5.1)

Let $f(x) = |x^4 + x - 1| - x$.

(a) Plot both f and f' on the same axes.

(b) Locate the critical points of f and estimate each local maximum and minimum correct to six decimal places.

(c) Find the global maximum and minimum values of f.

Solution to (a) First, define f in Maple; then use the `diff` command to get the derivative.

```
>f := x -> abs(x^4 + x - 1) - x;
```

$$f := x \to |x^4 + x - 1| - x$$

```
>df := diff(f(x), x);
```

$$df := \frac{|x^4 + x - 1|(4x^3 + 1)}{x^4 + x - 1} - 1 \qquad \text{(Release 2 and earlier)}$$

$$df := abs(1, x^4 + x - 1)(4x^3 + 1) - 1 \qquad \text{(Release 3)}$$

The result of the last command depends on which release of Maple you are working with. For Release 3 users, $abs(1, x^4 + x - 1)$ is synonymous with $\dfrac{|x^4 + x - 1|}{x^4 + x - 1}$. In either case, the derivative of f involves the factor, $\dfrac{|x^4 + x - 1|}{x^4 + x - 1}$. The function $\dfrac{|x|}{x}$ in Maple is called the `signum` function. It is -1 when x is negative, $+1$ when x is positive, and is undefined at $x = 0$. We already know that $|x|$ is not differentiable at $x = 0$, and in fact $\dfrac{d|x|}{dx} = \text{signum}(x)$. For Release 3 users, note that $abs(0,x) = |x|$, $abs(1,x) = \text{signum}(x)$, and $abs(2, x) = \dfrac{d}{dx} \text{signum}(x) = \text{signum}(1, x)$.

Since the derivative does not exist where the denominator is zero, we should expect some unusual behavior at these points. To get the graphs, enter

```
>plot({f(x), df}, x = -2..2, y = -20..20);
```

The result appears in Figure 5.2a. Notice that Maple does not show a jump in the graph of the derivative at about $x = -1.22$ and at about $x = -0.75$ as it should. Instead, it has drawn vertical line segments at these points. This is caused by the way the program makes graphs by connecting dots with line segments. We can demonstrate that these vertical line segments should not be there by selecting **point** from the **style** menu in the plot window. Maple will then redraw the graph using only points and the vertical lines will not appear. Alternatively, we can ask Maple to plot the graph directly as points and even tell it how many points to use by executing the following command; see Figure 5.2b.

```
>plot({f(x), df}, x = -2..2, y = -20..20, style = POINT, numpoints = 100);
```

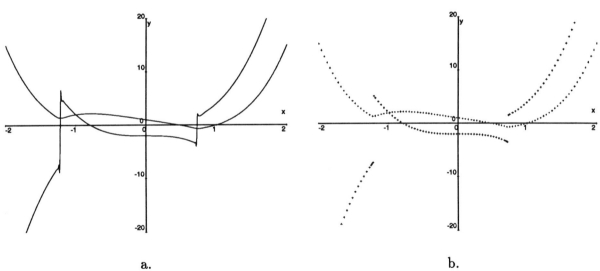

a.   b.

Figure 5.2: Plot of $f(x) = |x^4 + x - 1| - x$ and its derivative

Solution to (b): A *critical value* of f is a value of x where the derivative of f is zero or undefined. From the graph, we see three critical points clearly: one negative value where the graph of the derivative crosses the x-axis and two discontinuities where it is undefined. Using the mouse pointer and clicking, we can determine the approximate coordinates of the local maxima and minima. For example, at the negative discontinuity, we estimate the local minimum at $x = -1.22$ and $y = 1.25$. The other local extrema and critical values may be approximated in the same way. The *accuracy* of such graphical estimates is questionable. By zooming in, we improve the estimates, but to find the critical values to a known degree of accuracy, it is wise to use the derivative. From the graph of f' we know that the critical point where $f'(x) = 0$ occurs in the interval $[-1, 0]$. To find this critical point we use `fsolve` in this interval.

```
>fsolve(df = 0, x, -1..0);
```
$$-.7937005260$$

This value is accurate to 10 significant digits and so meets the condition of this Part.

The other two critical values occur where the derivative is undefined, that is, at the points where $x^4 + x - 1 = 0$. From the graph of the derivative, we see that this occurs on the intervals $[-2, -1]$ and $[0, 1]$. We therefore use `fsolve` again but this time on $x^4 + x - 1$.

```
>fsolve(x^4+x-1 = 0, x, -2..-1);
```
$$-1.220744085$$

```
>fsolve(x^4+x-1 = 0, x, 0..1);
```
$$.7244919590$$

By asking Maple to compute $f(-1.220744)$ and $f(0.724492)$, we obtain the corresponding y values. We conclude that there are local minima at $(-1.220744, 1.220745)$ and at $(0.724492, -0.724492)$, and a local maximum at $(-0.793701, 2.190551)$. (Compare these with the answers obtained graphically.)

Solution to (c): There is no global maximum, but $y = -0.724492$ is a global minimum occurring at $x = 0.724492$.

Laboratory Exercise 5.1

Local Extrema (CCH Text 5.1)

Name _____ Due Date _____

Find all the critical values of each function f below to six decimal place accuracy, and find all the local (or relative) extrema as well as absolute (or global) extrema of f. Explain how you arrive at each of your answers. Plot the graph of f and its derivative so that all local extrema are shown.

1. $f(x) = x^2 + \sin x$.

2. $f(x) = \dfrac{x^2 + 1}{x + |x^3 + x - 1|}$.

3. $f(x) = \sqrt{x^4+1} - \sqrt{x^2+1}$.

Solved Problem 5.3: Inflection points (CCH Text 5.2)

Plot the graphs of $f(x) = x^6 - x^4$ and its second derivative on the same axes. Find the inflection points of f.

Solution: First, we define f in Maple.

```
>f := x -> x^6 - x^4;
```
$$f := x \to x^6 - x^4$$

Normally, we would differentiate f twice to get the second derivative, but Maple has a feature that allows us to get the second derivative directly. We use the `diff` command together with the $ operator. The $ operator simply repeats the expression that precedes it the number of times we specify. Thus `x$2` means `x, x`. To get the second derivative of f, we can therefore either enter `diff(f(x), x$2)` or `diff(f(x), x, x)`. We choose the former calling the second derivative, *ddf*.

```
>ddf := diff(f(x), x$2);
```
$$ddf := 30x^4 - 12x^2$$

Next plot both f and *ddf* on the same axes using the following command. The result appears in Figure 5.3. Can you tell which graph is f and which is *ddf*?

```
>plot({f(x), ddf}, x = -2..2, y = -2..2);
```

Points of inflection *may* occur where the second derivative is zero. Hence we ask Maple to solve $ddf = 0$ for x.

```
>s := solve(ddf = 0, x);
```
$$s := 0,\ 0,\ \frac{1}{5}\sqrt{10},\ -\frac{1}{5}\sqrt{10}$$

Maple presents us with four solutions, two of which are 0, in an *expression sequence* which we have called s. This will make it easier for us to refer to the individual solutions.

The graph of the second derivative crosses the x-axis indicating a change in concavity in the graph of f at both $x = -\frac{\sqrt{10}}{5}$ and $x = \frac{\sqrt{10}}{5}$. Thus, these are the x coordinates of inflection points. On the other hand, since the graph of the second derivative does not cross the x-axis at $x = 0$, no change in concavity occurs there. We conclude that $(0,0)$ is not an inflection point.

To find the y coordinates of the inflection points, we ask Maple to find $f(\frac{\sqrt{10}}{5})$, and the symmetry of the graph tells us that the other inflection point has this same y coordinate. Note that $\frac{\sqrt{10}}{5}$ is the third term in the expression sequence s, so we can refer to it in Maple as `s[3]`.

```
>f(s[3]);
```
$$\frac{-12}{125}$$

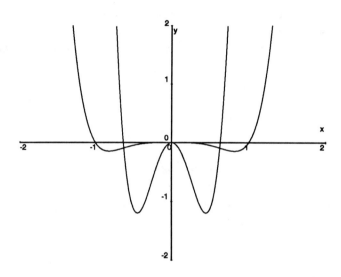

Figure 5.3: Inflection points for $x^6 - x^4$

Laboratory Exercise 5.2

Inflection Points (CCH Text 5.2)

Name _____   Due Date _____

For each of the following functions, plot the graphs of f and f'', and find all the inflection points of f.

1. $f(x) = \arctan(1 + x^2)$

2. $f(x) = x^2 e^{-x}$

3. $f(x) = x^4 \ln(1 + e^x)$

Laboratory Exercise 5.3

The George Deer Reserve (CCH Text 5.2)

Name _____   Due Date _____

The *logistic model of population growth* predicts that the population (as a function of time) of deer on the George Deer Reserve in Michigan is given by

$$P(t) = \frac{5.1}{0.03 + e^{-0.59t}}$$

1. Plot the graph of $P(t)$.

2. The *Maximum Sustainable Yield* model for wildlife management says that populations should be maintained at a level where population growth is a maximum. Express this statement in terms of inflection points of $P(t)$.

3. According to the Maximum Sustainable Yield model, at what level should the deer population on the George Deer Reserve be maintained?

4. What is the maximum population of deer that the George Deer Reserve can support? Explain clearly how you arrived at your answer.

Laboratory Exercise 5.4

The Meaning of the Signs of f' and f'' (CCH Text 5.2)

Name _____   Due Date _____

This is a continuation of Lab Exercise 2.6. It requires a color monitor. We will produce a graphical demonstration of the following important principles: If f' is positive, then f is increasing. If f' is negative, then f is decreasing. If f'' is positive, then f is concave up. If f'' is negative, then f is concave down.

1. First define f := x -> sin(x) in Maple. Next, at a Maple prompt, enter the following lines <u>exactly</u> as they appear here. Be sure to pay attention to the syntax, upper and lower case letters, and to press return after each line.

```
>p := proc(x) if D(f)(x) > 0 then f(x) else NOP fi; end;
>q := proc(x) if D(f)(x) < 0 then f(x) else NOP fi; end;
```

Once you have done this, enter the following and explain what you see.

```
>plot({p, q}, 0..2*Pi);
```

2. In Maple, D(f) is the notation for the derivative function, f', and D(f)(x) is just $f'(x)$. Explain what the formula in Part 1 is doing and how it works.

3. At a Maple prompt, enter the following <u>exactly</u> as they appear here. Be sure to pay attention to the syntax, upper and lower case letters, and to press return after each line.

```
>P := proc(x) if D(D(f))(x) > 0 then f(x) else NOP fi; end;
>Q := proc(x) if D(D(f))(x) < 0 then f(x) else NOP fi; end;
```

Once you have done this, enter the following and explain what you see.

```
>plot({P, Q}, 0..2*Pi);
```

4. In Maple, `D(D(f))` denotes the second derivative function, f'', and `D(D(f))(x)` is just $f''(x)$. Explain what the formula in Part 3 is doing and how it works.

5. Repeat Parts 1-4 above for $f(x) = x^3 - x$.

Plotting families of curves (CCH Text 5.3)

Maple's **seq** function (for *sequence*) is an ideal tool for plotting families of curves. There are two forms for the syntax of this command that can be useful to us. The first is used when we wish to generate a collection of functions in an orderly sequence. For example, if we wish to consider the functions, $\sin x$, $\sin 2x$, $\sin 3x$, and $\sin 4x$, Maple will produce this sequence for us as follows:

```
>seq(sin(k*x), k = 1..4);
```

$$\sin(x),\ \sin(2x),\ \sin(3x),\ \sin(4x)$$

To create a sequence of functions that is not so orderly, we first create a list containing the values of k we wish to consider. For example, suppose we wish to consider the collection, $\sin(0.5x)$, $\sin(2x)$, and $\sin(x)$. We create a list, (we'll call it L, but you may use any name you wish), which contains the values 0.5, 2, and 1:

```
>L := [0.5, 2, 1];
```

$$L := [.5, 2, 1]$$

Then we use the **seq** command again, but this time as follows:

```
>seq(sin(k*x), k = L);
```

$$\sin(.5x),\ \sin(x),\ \sin(2x)$$

The **seq** command is very useful. To use it to plot a family of functions, we enclose the command in set braces to have Maple generate a *set* of functions instead of a sequence of functions. For instance, in the above example such a set may be generated as follows. Note that here we give the set a name.

```
>s := {seq(sin(k*x), k = L)};
```

$$s := \{\sin(.5x), \sin(x), \sin(2x)\}$$

We can now plot this family in the usual way. The result of the following command appears in Figure 5.4.

```
>plot(s, x = -Pi..Pi);
```

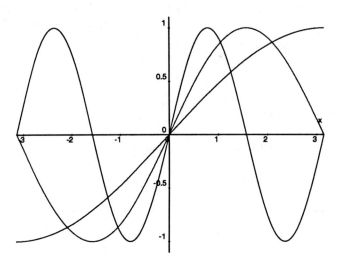

Figure 5.4: The family of curves, sin(kx)

Laboratory Exercise 5.5

Families of Curves (CCH Text 5.3)

Name _____   Due Date _____

1. Use the `seq` command as shown on p. 153 to plot $\sin(e^{kx})$ for at least six positive values of k. Clearly label each curve in the printout of your plot with the corresponding value of k.

2. Describe what happens to $\sin(e^{kx})$ for positive x as k gets larger. Describe what happens to $\sin(e^{kx})$ for negative x as k gets larger.

3. Use the `seq` command as shown on p. 153 to plot $x^2 + kx + 1$ for at least five positive values of k and five negative values. Clearly label each curve in the printout of your plot with the corresponding value of k.

4. Describe how the family of curves $x^2 + kx + 1$ changes with k.

5. The vertices of the family of graphs $x^2 + kx + 1$ appear to lie on a single parabola. Find a quadratic function that goes through all their vertices and plot it along with the family.

Solved Problem 5.4: Welding boxes (CCH Text 5.6)

A company has a contract to build several open metal trash bins. Each has a square base and will hold 1000 cubic feet. It orders a pre-cut sheet for the bottom of the box and another that it bends three times to form the four sides. (There is no top.) It must then weld the seams (one vertical and four horizontal). Records indicate that welding costs $2.10 per foot including labor and materials. The sheet metal costs $1.85 per square foot.

(a) What are the dimensions of the box that costs the least to build?

(b) What is the cost of the box described in Part (a)?

(c) Another company figures it will just make the box 10 feet on a side. How much more does it cost to do this?

Solution to (a): Let h be the height of the box and b the length (and width) of the base.

TOTAL COST = COST OF METAL + COST OF WELDING.

The bottom has area b^2 square feet, and there are four sides each having area bh square feet. Thus the total cost of the sheet metal used to build the box is $1.85(b^2 + 4bh)$. The total length to be welded is the perimeter of the base, $4b$ feet, plus the height of the seam along a side, h feet. Therefore the welding costs $2.10(4b+h)$. Adding these gives the total cost.

$$\text{COST} = 1.85(4bh + b^2) + 2.10(h + 4b)$$

Since $1000 = hb^2$, we get $h = \frac{1000}{b^2}$. Now we can express the cost in terms of a single variable, b.

$$\text{COST} = 1.85\left(b^2 + \frac{4000}{b}\right) + 2.10\left(4b + \frac{1000}{b^2}\right)$$

Define the cost function in Maple; then ask Maple to get its derivative, calling it dCost:

```
>Cost := 1.85*(b^2 + 4000/b) + 2.10*(4*b + 1000/b^2);
```

$$Cost := 1.85 b^2 + 7400.00\frac{1}{b} + 8.40 b + 2100.00\frac{1}{b^2}$$

```
>dCost := diff(Cost, b);
```

$$dCost := 3.70 b - 7400.00\frac{1}{b^2} + 8.40 - 4200.00\frac{1}{b^3}$$

A graph of the derivative appears in Figure 5.4. This graph shows that the derivative crosses the horizontal axis between 10 and 20. In order to get this critical value, we will use Maple's `fsolve` command in the interval $[10, 20]$ to solve the equation $dCost = 0$ for b.

```
>fsolve(dCost = 0, b, 10..20);
```
$$12.07976991$$

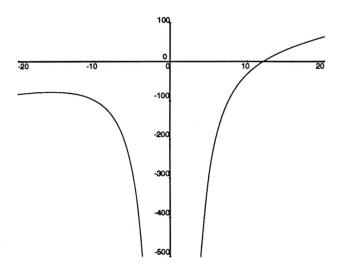

Figure 5.5: Derivative of the cost function

We see that the base should be 12.0797 feet on a side, and the height should be $\dfrac{1000}{12.0797^2} \approx 6.853$ feet.

Solution to (b): We now need to calculate how much the box costs. This is obtained by substituting the value found for b in Part (a) into the cost function. After executing the following command and rounding to two decimal places, we see that a box costs $998.41.

```
>subs(b = 12.0797, Cost);
```

Solution to (c): We need to calculate how much the box would cost if we made it 10 feet on each side. Execute the last command with 12.0797 replaced by 10. The result is $1030.00, $31.59 (or 3.2%) more than the cheapest way. So, the contractor who knows calculus may underbid her less astute competitor and win the contract.

Laboratory Exercise 5.6

Building Boxes (CCH Text 5.6)

Name _____   Due Date _____

A contractor wants to bid on an order to make 150,000 boxes out of cardboard that costs 12 cents per square foot. The base of each box must be square and reinforced with an extra layer of cardboard. The contractor must assemble each box by taping the four seams around the bottom, one seam up the side, and one seam on top to make a hinged lid. Company records indicate that taping costs 11 cents per foot including labor and materials.

1. Assuming that the boxes are to hold 3.5 cubic feet, write the cost of a single box as a function of the length of the base.

2. What should the dimensions be to ensure the lowest cost? Show your work.

3. If the contractor wants to make a profit of 17%, what should she bid?

4. A competitor, who also wants to make a profit of 17%, prepares a bid for boxes that are perfect cubes. By how many dollars does this competitor lose the bid?

Laboratory Exercise 5.7

Building Fuel Tanks (CCH Text 5.6)

Name _____   Due Date _____

A firm is asked to build a cylindrical fuel tank to hold 300 gallons. It orders a rectangular piece of metal which it will roll into a cylinder, and two circular pieces to make the ends. It must then weld the seam of the rolled metal sheet to form a cylinder and the seams to fasten the top and bottom to the cylinder. The rectangular piece must be made of a malleable alloy costing $1.25 per square foot. The circular ends can be a cheaper metal costing 75 cents per square foot. Company records indicate that welding costs 50 cents per foot including labor and materials.

1. Write the cost as a function of the radius of the base.

2. What should the dimensions be to ensure the lowest cost? (There are 7.5 gallons in a cubic foot.) Show your work.

3. After the cost is figured, the customer decides she can afford to pay only 80% of the manufacturer's minimum cost computed above. What is the maximum volume that can be produced for this amount?

Laboratory Exercise 5.8

Roads (CCH Text 5.6)

Name _____   Due Date _____

A road from City A to City B must cross a strip of private land and a strip of public land as shown in Figure 5.5 below. Due to fees demanded by the owner, the cost of building the road on private land is 20% more per mile than it is on public land. Assuming the cost on public land is $89,650 per mile, what is the minimum cost of the project? Draw a road map with distances labeled that ensures a minimum cost, and write an explanation that the City Commissioners can understand.

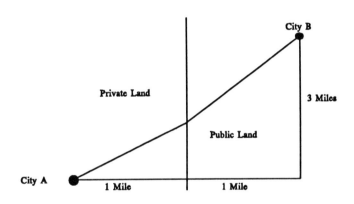

Figure 5.6: A road

Solved Problem 5.5: Newton's method (CCH Text 5.7)

Apply 4 steps of Newton's method to $f(x) = \sin x + \cos x$ beginning at $x_0 = 0.6$. Use the procedure NEWTONPIC provided in Appendix II to make a picture of this procedure.

Solution: The derivative of f is $\cos x - \sin x$. Thus the formula we need is:

$$x_{n+1} = x_n - \frac{\sin x_n + \cos x_n}{\cos x_n - \sin x_n}$$

If we take $NT(x)$ to be $x - \dfrac{\sin x + \cos x}{\cos x - \sin x}$, then $x_{n+1} = NT(x_n)$.

First, we define $NT(x)$ in Maple.

```
>NT := x -> x - (sin(x) + cos(x))/(cos(x) - sin(x));
```

$$NT := x \to x - \frac{\sin(x) + \cos(x)}{\cos(x) - \sin(x)}$$

To get the Newton iterates, we can evaluate $NT(0.6)$ calling it a; then we'll repeat the procedure on a four times. Alternatively, we can set $a = 0.6$ and then execute a `for` loop as follows:

```
>a := 0.6;
```

$$a := 0.6$$

```
>for n from 1 to 4 do a := NT(a); od;
```

$$a := -4.731855223$$
$$a := -3.692144707$$
$$a := -3.931405705$$
$$a := -3.926990788$$

The last value displayed is $x_4 = -3.926990788$.

To make a picture of all this requires us to type in some code that is provided in the section of Appendix II entitled **Numerical Routines: Newton's method**. After following the instructions to make the file, **Newton**, we read it into Maple's memory with the command,

>read 'Newton';

To get the graph, first define the function $f(x) = \sin x + \cos x$ in Maple with

>f := x -> sin(x) + cos(x);

$$f := x \to \sin(x) + \cos(x)$$

Now execute the following command which is part of the **Newton** file. The result is displayed in Figure 5.6.

>NEWTONPIC(f, 0.6, 4);

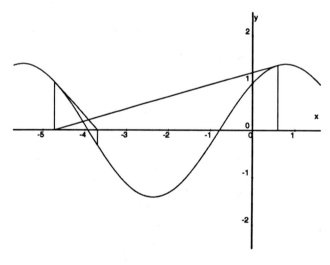

Figure 5.7: A picture of Newton's method

Laboratory Exercise 5.9

Newton's Method (CCH Text 5.7)

Name _____   Due Date _____

1. Apply five steps of Newton's method to $x^2 - \cos x$ and use the file **Newton** to make a picture of the procedure.

2. Make a picture of 10 iterations of Newton's method for $\sin x + \cos x$ for starting points 0.8 and 1.1. Explain what you observe.

3. Find a starting point for Newton's method applied to $\sin x + \cos x$ that will cause Newton's method to get caught in a loop. That is, $x_2 = x_0$. (Hint: If NT is the function used in Solved Problem 5.5, explain why a solution of $NT(NT(x)) = x$ will give the desired result.)

4. Produce a picture of Newton's method applied to $\sin x + \cos x$ using the starting point found in Part 3.

Chapter 6
Reconstructing a Function from Its Derivative

One of the primary goals of Chapter 6 in your CCH Text is to teach some elementary facts about integration. You may find Maple useful for checking your answers, but most of the work in this chapter is better done without a computer.

Solved Problem 6.1: Families of antiderivatives (CCH Text 6.3)

Let $f(x) = \ln x$. Plot a family of antiderivatives of f and find the antiderivative whose value is 3 when $x = 1$.

Solution: Begin by defining f in Maple:

>f := x -> ln(x);
$$f := \ln$$

Now use Maple's int (for *integrate*) command to integrate $f(x)$ with respect to x.

>int(f(x), x);
$$x \ln(x) - x$$

Thus the family of antiderivatives of $\ln x$ is $x \ln x - x + c$ where c can take any value we choose. Notice that Maple does not provide the arbitrary constant c. We must do that ourselves. One way to do this that will be useful is to use the expression $x \ln x - x + c$ to define a function of the constant c. We'll call this function C.

>C := c -> int(f(x), x) + c;
$$C := c \to \int f(x)\,dx + c$$

The response we get from Maple may appear to be strange but to verify that indeed we have the correct function, ask Maple for a value of C, say $C(1)$:

>C(1);
$$x \ln(x) - x + 1$$

Now that we know that we have the correct function, we will use Maple's `seq` function as we did on p. 153 to plot this family of curves. In the following we create a set of the antiderivatives $x \ln x - x + c$ as c ranges from -8 to 8.

```
>s := {seq(C(c), c = -8..8)};
```

The result of this command is a listing of the elements of s, some of which are $x \ln x - x + 8$, $x \ln x - x - 2$, $x \ln x - x + 5$, etc. Your screen will contain all 17 antiderivatives.

The graph shown in Figure 6.1 is the output of the following `plot` command.

```
>plot(s, x = 0..10, y = -10..10);
```

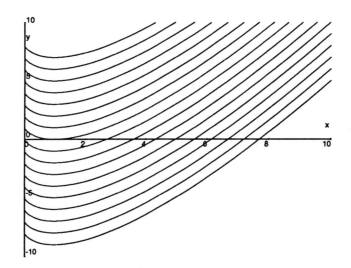

Figure 6.1: Antiderivatives of $\ln x$

In order to find the antiderivative that has the value 3 when $x = 1$, we solve the equation $1 \ln 1 - 1 + c = 3$ (by hand). This gives $c = 4$, so the antiderivative we seek is $x \ln x - x + 4$.

Laboratory Exercise 6.1

Antiderivatives of arctan x (CCH Text 6.3)

Name _____   Due Date _____

1. Plot a family of at least 10 antiderivatives of arctan x. Explain how the graphs are related.

2. Find an antiderivative of arctan x that has value 5 when $x = 1$ and plot its graph. Can there be more than one such antiderivative? Explain your answer.

Laboratory Exercise 6.2

An Antiderivative of $\sin(x^2)$ (CCH Text 6.3)

Name _____   Due Date _____

Let $f(x) = \sin(x^2)$. If you ask Maple to find the antiderivative of $f(x)$, it will return a rather complex function that would require more explanation than we wish to give at this level. However, we can still examine this antiderivative.

1. Define $g(x) = \int_0^x \sin(t^2)\, dt$; then verify that $g(x)$ is an antiderivative of $f(x)$ by asking Maple to find its derivative.

2. Plot the graph of $g(x)$. This may take some time depending on the speed of your computer.

3. What is the exact value of $g(0)$? Does any other antiderivative of $f(x)$ have this value at 0? Explain.

4. $g(x)$ is one of many antiderivatives of $f(x)$. Plot the graph of the antiderivative whose value is 3 when $x = 0$. Does any other antiderivative of $f(x)$ have the value 3 at $x = 0$? Explain.

Chapter 7
The Integral

We are familiar with approximating integrals using left-hand and right-hand Riemann sums. In this chapter these procedures are referred to as LEFT(n) and RIGHT(n), where n is the number of subintervals. Improved approximation schemes are also presented: The *midpoint rule:* MID(n), the *trapezoidal rule:* TRAP(n), and *Simpson's rule:* SIMP(n).

We will help you produce Maple files (which you can save for later use) that implement these procedures. RIGHT(n) and MID(n) are implemented in Solved Problems 7.2 and 7.3, and the rest are implemented in Lab Exercises 7.2 and 7.3. All five approximation schemes have geometric interpretations that can be represented graphically. Maple provides procedures for some of these, and in Appendix II we provide the code for the others.

Maple also will find the antiderivatives of a great many functions and thus provides a convenient way to check your answers to exercises. However, it will not tell you *how* it gets its answers.

Solved Problem 7.1: Integrating with Maple (CCH Text 7.1)

(a) Find $\int \sin \sqrt{x}\, dx$ and check your answer.

(b) Find $\int \sin(x^2)\, dx$.

(c) Approximate $\int_0^1 \sin(x^2)\, dx$.

(d) Approximate $\int_0^1 \sin(x^3)\, dx$.

Solution to (a): The following command will evaluate the integral in this Part.

>int(sin(sqrt(x)), x);
$$2\sin(\sqrt{x}) - 2\sqrt{x}\cos(\sqrt{x})$$

To check the integral, we differentiate the above using `diff`.

```
>diff(", x)
```
$$\sin(\sqrt{x})$$

Solution to (b): This example was the subject of Lab Exercise 6.2. As mentioned there, Maple will integrate this function, but the result is not expressed in terms of the usual "simple" functions.

```
>int(sin(x^2), x);
```
$$\frac{1}{2}\sqrt{2}\sqrt{\pi}\ \text{FresnelS}\left(\frac{\sqrt{2}x}{\sqrt{\pi}}\right)$$

Solution to (c): Even though we do not understand the function Maple has produced, we can still approximate the definite integral. To do so, first ask Maple to evaluate it.

```
>int(sin(x^2), x = 0..1);
```
$$\frac{1}{2}\sqrt{2}\sqrt{\pi}\ \text{FresnelS}\left(\frac{\sqrt{2}}{\sqrt{\pi}}\right)$$

Maple has produced the answer symbolically, meaning it has not given us a numerical value. Using **evalf**, we can get the numerical approximation.

```
>evalf(");
```
$$.3102683014$$

Solution to (d): When we ask Maple to evaluate this integral, it will "think" about it for a while, and then return the integral to us. This is Maple's way of saying that it cannot find the answer.

```
>int(sin(x^3), x = 0..1);
```
$$\int_0^1 \sin(x^3)\,dx$$

Even though Maple cannot find the integral symbolically as it did in Part (c), it can still approximate the definite integral. This may be done as in the solution to Part (c).

```
>evalf(");
```
$$.2338452456$$

Laboratory Exercise 7.1

Integrating with Maple (CCH Text 7.1)

Name _____   Due Date _____

1. Use Maple to find $\int \sec^2 x \, dx$.

2. Based on your answer in Part 1, what should be the derivative of $\dfrac{\sin x}{\cos x}$? Note that $\tan x = \dfrac{\sin x}{\cos x}$. What is the antiderivative of $\tan x$?

3. Use Maple to find the derivative of $\tan x$. Reconcile the answer here with your answer in Part 2.

Solved Problem 7.2: Implementing the right-hand rule (CCH Text 7.6)

Use Maple's `rightsum` function in the `student calculus` package to approximate the integral, $\int_1^3 x^2\, dx$ using 20 subintervals.

Solution: We begin by reading the `student calculus` package into Maple's memory.

`>with(student);`

(See Solved Problem 3.2.) When this command is executed, Maple will read the calculus package into its memory and present us with a listing of the now available routines. The ones we are interested in here are `rightsum` and `value`.

The syntax for `rightsum` is as follows:

$$\text{rightsum(expr, x = a..b, n)}$$

where `expr` is the Maple expression whose Riemann sum we seek, `x = a..b` is the interval over which we wish to approximate the integral of the expression, and `n` is the optional number of subintervals. If `n` is not provided, it defaults to 5. The result of `rightsum` is the right-hand Riemann sum, $\sum_{i=1}^{n} f(x_i)\Delta x$ where $\Delta x = \dfrac{(b-a)}{n}$ and $x_i = a + i\Delta x$. Thus to approximate the given integral, we start by entering

`>rightsum(x^2, x = 1..3, 20);`

$$\frac{1}{10}\left(\sum_{i=1}^{20}\left(1+\frac{1}{10}i\right)^2\right)$$

Notice that Maple produced the *symbolic* right-hand Riemann sum by actually evaluating $f(a + i\Delta x)$ when $f(x) = x^2$. To get the value of this sum, we use the `value` command:

`>value(");`

$$\frac{907}{100}$$

The exact value of the integral is $\frac{26}{3} \approx 8.66666$. Since x^2 is an increasing function, the right-hand Riemann sum overestimates the integral and the value 9.07 we obtained from Maple is larger than the true value as expected.

Solved Problem 7.3: Implementing the midpoint rule (CCH Text 7.6)

(a) Use the `middlesum` function with 20 subintervals to approximate $\int_1^3 x^2 \, dx$.

(b) Use the `middlebox` function to create a picture representing the midpoint sum for 10 subintervals.

Solution to (a): If it has not already been entered in Maple's memory, read the student calculus package in now as we did in Solved Problem 7.2. The routines we will be interested in here are `middlesum`, `middlebox`, and `value`.

The syntax for `middlesum` is exactly the same as that for `rightsum`. Using `middlesum` we get the Riemann sum, $\sum_{i=0}^{n-1} f(x_i)\Delta x$, where $\Delta x = \dfrac{(b-a)}{n}$ and $x_i = a + \left(i + \dfrac{1}{2}\right)\Delta x$. (We leave it as an exercise for the reader to show that this formula for x_i gives the midpoint of the ith subinterval).

We proceed in much the same way as we did in Solved Problem 7.2. We begin by entering

```
>middlesum(x^2, x = 1..3, 20);
```

$$\frac{1}{10}\left(\sum_{i=0}^{19}\left(\frac{21}{20} + \frac{1}{10}i\right)^2\right)$$

We again see that Maple has produced the symbolic middle sum. We get its value as before using Maple's `value` command.

```
>value(");
```

$$\frac{1733}{200}$$

We can also get the result as a decimal, 8.665, using `evalf`. Notice that this answer is much closer to the exact value, $\frac{26}{3}$, of the integral than we got using the `rightsum` function. Furthermore, x^2 is concave up, so, as expected, `middlesum` gave an underestimate of the definite integral.

Solution to (b): You will also notice that in the student calculus package there is a routine called `middlebox`. We use it here to produce a picture of the midpoint sum by entering the following command whose picture appears in Figure 7.1.

```
>middlebox(x^2, x = 1..3, 10);
```

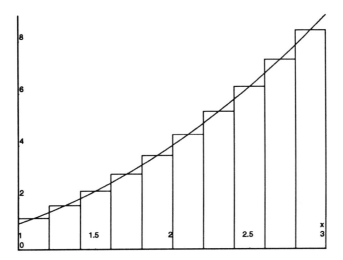

Figure 7.1: Graphical representation of the midpoint rule

Laboratory Exercise 7.2

Implementing the Left-Hand Rule (CCH Text 7.6)

Name _____   Due Date _____

1. Use the `leftsum` function in Maple's `student calculus` package with 10 subintervals to approximate $\int_0^{\frac{\pi}{2}} \sin x \, dx$.

2. Use the `rightsum` function with 10 subintervals to approximate $\int_0^{\frac{\pi}{2}} \sin x \, dx$.

3. Discuss how the approximations in Parts 2 and 3 compare to the exact value of $\int_0^{\frac{\pi}{2}} \sin x \, dx$.

4. Use the `leftbox` and `rightbox` routines in the **student calculus** package to create pictures representing the left and right sums in Parts 2 and 3. Explain the relationship between the pictures and the approximations.

5. Use the `leftsum` function with 10 subintervals to approximate $\int_{\frac{\pi}{2}}^{\pi} \sin x \, dx$.

6. Use the `rightsum` function with 10 subintervals to approximate $\int_{\frac{\pi}{2}}^{\pi} \sin x \, dx$.

7. Discuss how the approximations in Parts 5 and 6 compare to the exact value of $\int_{\frac{\pi}{2}}^{\pi} \sin x \, dx$.

8. Use the `leftbox` and `rightbox` routines in the `student calculus` package to create pictures representing the left and right sums in Parts 5 and 6. Explain the relationship between the pictures and the approximations.

9. Discuss the difference between what you observed in Parts 3 and 7. What caused the difference?

Laboratory Exercise 7.3

Implementing the Trapezoidal Rule and Simpson's Rule
(CCH Text 7.7)

Name _____   Due Date _____

1. Use the `trapezoid` function in Maple's `student calculus` package with 10 subintervals to approximate $\int_0^{\pi/2} \sin x \, dx$. The `trapezoid` function implements the trapezoidal rule.

2. Use the TRAPBOX routine given in Appendix II to create a picture representing the trapezoidal rule in Part 1. Explain the relationship between the picture and the approximation.

3. Use the `simpson` function in Maple's `student calculus` package with 10 subintervals to approximate $\int_0^{\pi/2} \sin x \, dx$. The `simpson` function implements Simpson's Rule.

4. Use the SIMPBOX routine given in Appendix II to create a picture representing Simpson's Rule in Part 3. Explain the relationship between the picture and the approximation.

5. Discuss how the various approximations (LEFT, RIGHT, MID, TRAP, and SIMP) compare to the exact value of $\int_0^{\frac{\pi}{2}} \sin x \, dx$. Observe which of the approximations is an overestimate and which is an underestimate. Which approximation is best and which is worst?

Laboratory Exercise 7.4

The Trapezoidal Rule with Error Control (CCH Text 7.7)

Name _____   Due Date _____

A note on the error in using the trapezoidal rule

We noted in Chapter 3 that if f is monotone (increasing or decreasing) on the interval $[a, b]$, then the error in approximating $\int_a^b f(x)\,dx$ using LEFT(n) or RIGHT(n) is no more than $|f(b) - f(a)|\frac{(b-a)}{n}$, and that using their average cuts the error in half. Since TRAP(n) is in fact the average of LEFT(n) and RIGHT(n), we may conclude:

If f is a monotone function on $[a, b]$, then the maximum error in using the trapezoidal rule, **TRAP(n)**, to approximate $\int_a^b f(x)\,dx$ is less than or equal to

$$|f(b) - f(a)|\frac{b-a}{2n}$$

---

1. What value of n is needed if TRAP(n) is to estimate $\int_0^1 \frac{4}{1+x^2}\,dx$ with error less than 0.01?

2. Calculate TRAP(n) for the value of n you found in Part 1.

3. Find the exact value of $\int_0^1 \frac{4}{1+x^2}\,dx$.

4. Obtain Maple's approximation for π to 20 significant digits. (Use `evalf(Pi, 20)`.) How close to π is your answer in Part 2?

5. How large would you need to choose n so that TRAP(n) gives an approximation that is accurate to 20 decimal places? Discuss the practicality of actually making this calculation.

Solved Problem 7.4: Calculating improper integrals (CCH Text 7.8)

Use Maple to calculate the following improper integrals. If the integral exists, verify that it is a limit of a Riemann integral.

(a) $\int_0^\infty x^4 e^{-x} \, dx$

(b) $\int_0^2 \frac{1}{(1-t)^2} \, dt$

(c) $\int_0^\infty \sin^3 x \, dx$

Solution to (a): Maple has the ability to evaluate many improper integrals, if possible, using its `int` command. All we have to do is to supply the needed limits of integration.

```
>int(x^4*exp(-x), x = 0..infinity);
                           24
```

Thus $\int_0^\infty x^4 e^{-x} \, dx$ converges and equals 24. The definition tells us that $\int_0^\infty x^4 e^{-x} \, dx = \lim_{k \to \infty} \int_0^k x^4 e^{-x} \, dx$. We can verify this with the following nested command.

```
>limit(int(x^4*exp(-x), x = 0..k), k = infinity);
                           24
```

Solution to (b): This one seems easy enough. Let's try this one without Maple's help. The integrand is $f(t) = \frac{1}{(1-t)^2}$ and an antiderivative of f may be seen to be $F(t) = \frac{1}{1-t}$. Therefore, $F(2) - F(0) = -2$, and so we conclude that $\int_0^2 \frac{1}{(1-t)^2} \, dt = -2$. But wait! The integrand $f(t) = \frac{1}{(1-t)^2}$ is positive, so how can the integral be negative? Let's check this with Maple.

```
>int(1/(1-t)^2, t = 0..2);
                           ∞
```

Infinity? What's going on? We notice that f has a vertical asymptote at $t = 1$ (we leave it to the reader to plot it), so this is an improper integral and must be checked for convergence. Our integral exists provided *both* the integrals $\int_0^1 \frac{1}{(1-t)^2} \, dt$ and $\int_1^2 \frac{1}{(1-t)^2} \, dt$ converge. Let's try the first integral in Maple.

```
>int(1/(1-t)^2, t = 0..1);
```
$$\infty$$

This shows that $\int_0^1 \frac{1}{(1-t)^2} \, dt$ does not exist, and hence that $\int_0^2 \frac{1}{(1-t)^2} \, dt$ diverges. Maple was trying to tell us this when we asked it to find $\int_0^2 \frac{1}{(1-t)^2} \, dt$. But the moral should be clear: Computer output must always be viewed critically.

Solution to (c): The integral here clearly is improper, so let's try it in Maple.

```
>int(sin(x)^3, x = 0..infinity);
```
$$undefined$$

Maple tells us that this integral is undefined. Being skeptical of computer output we can resort to other means to determine whether this integral converges or if indeed it is undefined. We try applying the definition of improper integral. Thus, if the limit exists, then $\int_0^\infty \sin^3 x \, dx = \lim_{k \to \infty} \int_0^k \sin^3 x \, dx$. We try this approach in the next two steps.

```
>int(sin(x)^3, x = 0..k);
```
$$-\frac{1}{3} \sin(k)^2 \cos(k) - \frac{2}{3} \cos(k) + \frac{2}{3}$$

```
>limit(", k = infinity);
```
$$-\frac{1}{3} .. \frac{4}{3}$$

We obtained a similar response from Maple in Solved Problem 2.5 when we tried to find $\lim_{x \to 0} \sin(1/x)$. This response from Maple seems to indicate the presence of oscillations, so perhaps a graph will help. Figure 7.2 is a plot of the definite integral given above. From this graph, we see that the antiderivative appears to oscillate back and forth between 0 and about 1.25. We conclude that the limit does not exist and hence $\int_0^\infty \sin^3 x \, dx$ diverges.

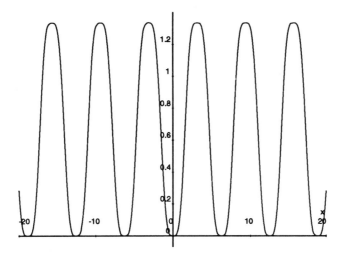

Figure 7.2: Divergence of $\int_0^\infty \sin^3 x\, dx$

Laboratory Exercise 7.5

Evaluating Improper Integrals (CCH Text 7.8)

Name _____   Due Date _____

Calculate the following improper integrals. Verify your answers by calculating the appropriate limit. Include any graphs you may have used to help you.

1. $\displaystyle\int_0^\infty x^3 2^{-x}\, dx$

2. $\displaystyle\int_0^1 \ln(x)\, dx$

3. $\displaystyle\int_0^4 \frac{1}{x^2-2}\, dx$

4. $\displaystyle\int_0^\infty \cos^5 x\, dx$

5. $\displaystyle\int_1^\infty \frac{1}{\arctan x}\,dx$

6. $\displaystyle\int_2^\infty \frac{1}{t^2+3t-4}\,dt$

7. $\displaystyle\int_1^2 \frac{1}{t^2+3t-4}\,dt$

8. $\displaystyle\int_1^\infty \frac{1}{t^2+3t-4}\,dt$

Laboratory Exercise 7.6

The Gamma Function Revisited (CCH Text 7.8)

Name _____   Due Date _____

In Lab Exercise 2.5, we encountered the **Gamma** function, but we were not told how it was defined. It is defined as the improper integral $\Gamma(x) = \int_0^\infty t^{x-1} e^{-t}\, dt$ and it is important in several areas of mathematics. In Lab Exercise 2.5 we estimated $\Gamma'(1)$ in several ways. In this exercise, we will try yet another.

1. Find $\dfrac{d}{dx} t^{x-1} e^{-t}$; then substitute $x = 1$ into the derivative.

2. Determine whether the integral of the result in Part 1 from 0 to ∞ exists. If so, what is its value? If not, explain why it doesn't.

3. Recall from Lab Exercise 2.5 that $\Gamma'(1) = -\gamma = -.5772156649$. Reconcile this fact with your answer to Part 2.

What you did in this lab is called "differentiating under the integral sign." It often works, but it must be justified.

Solved Problem 7.5: Approximating improper integrals (CCH Text 7.9)

Determine whether each of the integrals converges or diverges. If it converges, find or estimate its value.

(a) $\int_4^\infty \frac{1}{\ln t} \, dt$

(b) $\int_1^\infty \frac{1}{t^5 + t + 3} \, dt$

Solution to (a): We ask Maple to compute the integral with

```
>int(1/ln(t), t = 4..infinity);
```

Maple returns an error message that looks rather cryptic. When such a message occurs, it means that Maple has had some difficulty determining the result. In such a case, we can try the definition of an improper integral. To do this, we use the following nested command.

```
>limit(int(1/ln(t), t = 0..k), k = infinity);
```

Once again Maple's response is strange. It would be wiser at this point to abandon this approach and try another. Let's begin by plotting $\frac{1}{\ln t}$. See Figure 7.3a for the result of the following command.

```
>plot(1/ln(t), t = 0..5, y = -10..10);
```

Notice that the part of the graph of $\frac{1}{\ln t}$ that lies to the right of $t = 1$ in the first quadrant looks a lot like $\frac{1}{t}$. Let's plot $\frac{1}{t}$ on the same coordinate axes with $\frac{1}{\ln t}$. See Figure 7.3b.

```
>plot({1/t, 1/ln(t)}, t = 0..5, y = -10..10);
```

Notice that the graph of $\frac{1}{t}$ lies below the graph of $\frac{1}{\ln t}$ in the first quadrant, at least for $t > 1$. We conclude from these graphs that $\frac{1}{\ln t} > \frac{1}{t}$ for $t > 1$. (You should be able to show that this is always true. How does $\ln t$ compare with t?)

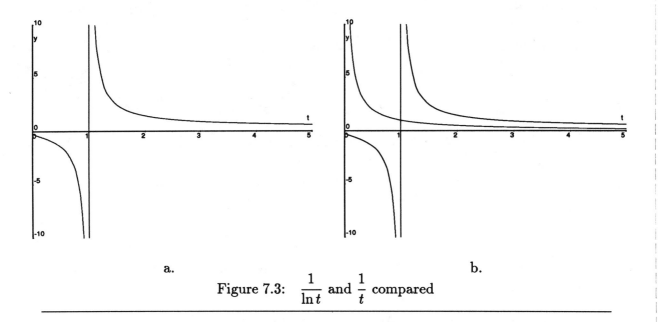

Figure 7.3: $\dfrac{1}{\ln t}$ and $\dfrac{1}{t}$ compared

We conclude that the two integrals are related as follows.

$$\int_4^\infty \frac{1}{\ln t}\, dt \geq \int_4^\infty \frac{1}{t}\, dt$$

But $\int_4^\infty \dfrac{1}{t}\, dt$ diverges. Thus, by the Comparison Test, $\int_4^\infty \dfrac{1}{\ln t}\, dt$ must also diverge. The lesson to be learned here is this: be sure to test for convergence before asking Maple for an approximation.

The approach we used here seems easy enough, why couldn't Maple do this? The answer is Maple can, but it would require us to delve a little deeper into Maple than we care to do so here.

Remark: If you are using Release 1 of Maple V, you already know this because that version returns "infinity" as the result from the first integral in the solution to Part (a).

Solution to (b): We'll apply the lesson we learned in Part (a) by checking for convergence before asking for an approximation. For $t \geq 0$, $t^5 + t + 3 > t^5 \geq 0$, so $\dfrac{1}{t^5 + t + 3} < \dfrac{1}{t^5}$. Hence, $\int_1^\infty \dfrac{1}{t^5 + t + 3}\, dt \leq \int_1^\infty \dfrac{1}{t^5}\, dt$. We know that the latter integral converges. Therefore, by the Comparison Test, $\int_1^\infty \frac{1}{t^5 + t + 3} dt$ must also converge.

We can now feel comfortable asking Maple for an approximation of the integral.

```
>int(1/(t^5 + t + 3), t = 1..infinity);
```

$$-\left(\sum_{\_R=\%1} \_R \ln\left(\frac{111476}{1875}\_R^2 + \frac{17131}{16875} - \frac{16216384}{16875}\_R^4 + \frac{4054096}{16875}\_R^3 + \frac{255941}{16875}\_R\right)\right)$$

$$\%1 := \quad \text{RootOf}(253381\_Z^5 - 160\_Z^3 - 80\_Z^2 - 15\_Z - 1)$$

Don't you feel more comfortable? Do not be put off by what you see here. Just keep in mind that Maple attempts to compute its result symbolically and that is exactly what it has done here. We are interested in a numerical approximation, so we ask Maple for it using **evalf**:

```
>evalf(");
```

$$.1087402244$$

How accurate is this answer? See Solved Problem 7.6.

Laboratory Exercise 7.7

Approximating Improper Integrals (CCH Text 7.9)

Name _____   Due Date _____

Maple either cannot evaluate the following improper integrals exactly or will return results that you may not understand. Determine if each integral converges. (A suggested comparison is given for each. You should analyze the comparison both graphically and algebraically.) For those integrals that converge, report Maple's approximation.

1. $\int_{2}^{\infty} \dfrac{1}{x^2 \ln x}\, dx$ (Suggested comparison: $\int_{2}^{\infty} \dfrac{1}{x^2}\, dx$)

2. $\int_{2}^{\infty} \dfrac{1}{\ln^2 t}\, dt$ (Suggested comparison: $\int_{2}^{\infty} \dfrac{1}{t \ln t}\, dt$)

3. $\int_0^\infty \dfrac{1}{x+e^x}\,dx$ (Suggested comparison: $\int_0^\infty e^{-x}\,dx$)

4. $\int_0^1 \dfrac{1}{x^3+\sqrt{x}}\,dx$ (Suggested comparison: $\int_0^1 x^{-\frac{1}{2}}\,dx$)

Solved Problem 7.6: Approximating improper integrals II (CCH Text 7.9)

In Solved Problem 7.5 we showed that $\int_1^\infty \frac{1}{t^5+t+3}\,dt$ converges and approximated it. The goal here is to get an approximation to a *known degree of accuracy*.

(a) Find a value of k so that $\int_1^k \frac{1}{t^5+t+3}\,dt$ approximates $\int_1^\infty \frac{1}{t^5+t+3}\,dt$ to two decimal places.

(b) Find an approximation to $\int_1^\infty \frac{1}{t^5+t+3}\,dt$ that you know is accurate to one decimal place.

Solution to (a): We will not use a computer here. For $\int_1^k \frac{1}{t^5+t+3}\,dt$ to approximate $\int_1^\infty \frac{1}{t^5+t+3}\,dt$ to two decimal places, we must have:

$$\int_1^\infty \frac{1}{t^5+t+3}\,dt - \int_1^k \frac{1}{t^5+t+3}\,dt < 0.005.$$

But $\int_1^\infty \frac{1}{t^5+t+3}\,dt - \int_1^k \frac{1}{t^5+t+3}\,dt = \int_k^\infty \frac{1}{t^5+t+3}\,dt.$ (Why?) Thus we need

$$\int_k^\infty \frac{1}{t^5+t+3}\,dt < 0.005.$$

Using the fact that $\frac{1}{t^5+t+3} < \frac{1}{t^5}$, we can conclude that for any $k > 0$

$$\int_k^\infty \frac{1}{t^5+t+3}\,dt < \int_k^\infty \frac{1}{t^5}\,dt = \frac{1}{4k^4}.$$

Therefore, if $\frac{1}{4k^4} < 0.005$, then certainly $\int_k^\infty \frac{1}{t^5+t+3}\,dt < 0.005$. Now $\frac{1}{4k^4} < 0.005$ implies that $k > \frac{1}{\sqrt[4]{4(0.005)}} \approx 2.65915.$

Taking $k = 3$, we can conclude that $\int_1^3 \frac{1}{t^5+t+3}\,dt$ approximates $\int_1^\infty \frac{1}{t^5+t+3}\,dt$ to within two decimal places.

Solution to (b): From our reasoning in Part (a), we know that $\int_1^3 \frac{1}{t^5 + t + 3} dt$ approximates $\int_1^\infty \frac{1}{t^5 + t + 3} dt$ to within two decimal places. Let's try this in Maple.

```
>int(1/(t^5 + t + 3), t = 1..3);
```

The result of this command is similar to the output we received in Solved Problem 7.5 and appears to be quite complex. Again, do not be daunted by what you see. Just **combine** the result, and then use **evalf** to get the approximation, 0.1056892215. Notice how close this value is to the approximation 0.1087402244 of $\int_1^\infty \frac{1}{t^5 + t + 3} dt$ that we obtained in Solved Problem 7.5. Both round to the same two decimal places as predicted!

These results give us confidence that Maple's approximations are quite accurate, but output like that from the above command may at times put us off, leaving us somewhat perplexed as to what to do next. Is there another way to approximate $\int_1^\infty \frac{1}{t^5 + t + 3} dt$ so that we *know* what the error is?

We will use **TRAP** in the **SUMS** file described in Appendix II. Make sure you read this file into Maple's memory before proceeding. To ensure two-place accuracy, we need to have

$$|f(3) - f(1)| \frac{3-1}{2n} \leq 0.005$$

where $f(t) = \frac{1}{t^5 + t + 3}$.

This gives $n > 39.1967$, so we will take $n = 40$ in **TRAP**. Now enter

```
>TRAP(1/(t^5 + t + 3), t = 1..3, 40);
```

$$.1057378687$$

We *know* that 0.1057378687 approximates $\int_1^3 \frac{1}{t^5 + t + 3} dt$ to two decimal places and that $\int_1^3 \frac{1}{t^5 + t + 3} dt$ approximates $\int_1^\infty \frac{1}{t^5 + t + 3} dt$ to two decimal places. Thus we can be sure that 0.1057378687 approximates $\int_1^\infty \frac{1}{t^5 + t + 3} dt$ to one decimal place. (Explain why.)

Even though Maple gives us very good approximations in a variety of computations, numerical procedures such as the trapezoidal rule or Simpson's Rule allow us to demonstrate a *definite* level of accuracy.

Laboratory Exercise 7.8

Approximating Improper Integrals II (CCH Text 7.9)

Name _____   Due Date _____

1. Show that $\displaystyle\int_1^\infty \frac{1}{t^6+t+1}\,dt$ converges.

2. Find a value of k so that $\displaystyle\int_1^k \frac{1}{t^6+t+1}\,dt$ approximates $\displaystyle\int_1^\infty \frac{1}{t^6+t+1}\,dt$ to two decimal places. Explain.

3. Find an approximation of $\int_1^\infty \frac{1}{t^6+t+1}\,dt$ that you are certain is accurate to one decimal place. Explain.

Chapter 8
Using the Definite Integral

This chapter looks at applications of the definite integral to geometry and physics. Maple can perform the calculations, but it's up to you to set up the problems.

Solved Problem 8.1: An oil slick (CCH Text 8.1)

Oil has spilled into a straight river 10 meters wide and has drifted downstream. The density of the oil is given by $d(x) = \dfrac{100}{1+x+x^2}$ kilograms per square meter, where $x \geq 0$ is the number of meters downstream from the source of the spill. (See Figure 8.1.) We assume that the density of the slick does not vary from one shore to the other.

(a) How much oil is within 25 meters of the source of the spill?

(b) How much oil was spilled into the river?

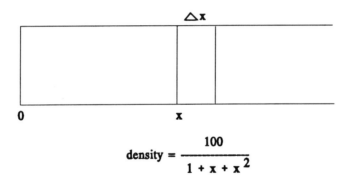

Figure 8.1: An oil spill in a straight river

Solution to (a): Since the density of the slick does not vary from one shore to the other, the amount of oil in a slice of the river of width Δx is:

$$\text{Density} \quad \times \quad \text{Area} \quad = \quad d(x) 10 \Delta x \qquad \text{(See Figure 8.1.)}$$

Thus the mass of the oil in a stretch of the river is approximated by a Riemann sum of the form $\sum d(x_i) 10 \Delta x$. In the 25 meters downstream from the source, the mass is given by $\displaystyle\int_0^{25} \dfrac{1000}{1+x+x^2}\, dx.$

Keeping in mind that Maple will evaluate the integral symbolically and that we are interested only in a numerical value, we enter the following nested command:

>evalf(int(1000/(1 + x + x^2), x = 0..25));

$$1169.998957$$

Thus, there are approximately 1170 kilograms of oil in the 25-meter stretch.

Solution to (b): The total oil spilled is the total mass of oil downstream. We weren't given the extent of the spill, so we will calculate the total amount of oil that would be in a slick of infinite length. That is, $\int_0^\infty \frac{1000}{1+x+x^2}\,dx$. This improper integral is easily evaluated by Maple.

>int(1000/(1+ x + x^2), x = 0..infinity);

$$\frac{2000}{9}\sqrt{3}\pi$$

>evalf(");

$$1209.199577$$

As a rough check we calculate the total mass of oil in a 1000-meter (or 1-kilometer or about 0.6-mile) slick:

>evalf(int(1000/(1 + x + x^2), x = 0..1000));

$$1208.200076$$

Thus, there is less than one kilogram more in a river of infinite length. Therefore it makes sense to assume the slick is infinite in length.

Solved Problem 8.2: Area and center of mass (CCH Text 8.1)

A thin sheet of metal is shaped like the region in the first quadrant between $y = \sin x$ and $y = \frac{x}{2}$. (See the region in Figure 8.2.)

(a) Find the area of the region.

(b) Find the x-coordinate of the center of mass of the region.

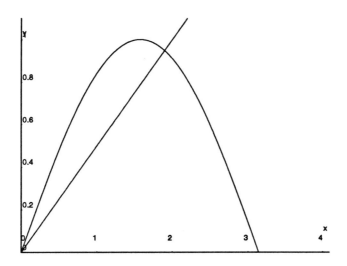

Figure 8.2: Area and center of mass

Solution to (a): Figure 8.2 shows the result of the following `plot` command.

```
>plot({sin(x), x/2}, x = 0..1.3*Pi, y = 0..1.1);
```

From the graph, we see that to find the area we will have to integrate $\sin(x) - \frac{x}{2}$ from $x = 0$ to the point where the graphs cross to the right of $x = 0$. Therefore we need to solve $\sin x = \frac{x}{2}$ for x. Maple will do this for us using `fsolve` on the interval $[1, 3]$.

```
>fsolve(sin(x) = x/2, x, 1..3);
```

$$1.895494267$$

which we round to six significant figures: 1.89549. Thus, the area is $\int_0^{1.89549} \sin x - \frac{x}{2}\, dx$. We calculate this area and call it A.

```
>A := int(sin(x) - x/2, x = 0..1.89549);
```

$$A := .4207978951$$

Therefore the area of the region to six significant figures is $A = 0.420798$.

Solution to (b): In general, the x-coordinate of the center of mass of the region bounded above by $y = f(x)$ and below by $y = g(x)$ on the interval $[a, b]$ is given by the formula

$$\frac{\int_a^b x(f(x) - g(x))\, dx}{\int_a^b f(x) - g(x)\, dx}$$

The numerator is called the "first moment" of the region about the y-axis. The denominator is the area of the region. If we placed this region on a thin rod parallel to the y-axis at this x-value, the region would balance on the rod. (See problem 2 in Section 8.1 of the CCH Text for a related exercise.)

To finish this problem, we need to find $\int_0^{1.89549} x(\sin x - \frac{x}{2})\, dx$ and divide it by the area found in Part (a). The first moment is

```
>M := int(x*(sin(x) - x), x = 0..1.89549);
```

$$M := .417399409$$

We can get the x-coordinate by computing M/A:

```
>M/A;
```

$$.9919237094$$

Therefore the x-coordinate of the center of mass to six significant figures is $x = 0.991924$. If the line, $x = 0.991924$, were a thin rod, the region would balance on the rod.

Laboratory Exercise 8.1

An Oil Spill in the Ocean (CCH Text 8.1)

Name _____   Due Date _____

A tanker ship illegally flushes its bilges off the Port of Galveston, making a circular oil slick with the ship at its center. The density of the slick x miles from the tanker is $\dfrac{8000}{1+e^x}$ gallons per square mile.

1. Set up a Riemann sum that approximates the total amount of oil in the slick when its radius is 8 miles. (<u>Hint</u>: Example 5 in Section 8.1 of the CCH Text offers a useful model for this problem.)

2. Find the amount of oil in the 8-mile slick.

3. Change the limits on your integral in Part 2 to 0 and ∞. Evaluate this new integral, and explain in practical terms what this number means.

Laboratory Exercise 8.2

The Center of Mass of a Sculpture (CCH Text 8.1)

Name _____   Due Date _____

    An art student was given a circular metal disk 2 feet in diameter and told to drill a small hole in it so that when the disk was cut in half and the piece with the hole was placed atop a spike stuck in the hole, it would balance. Not knowing about integration, the artist drilled a hole at a point halfway between the center and the edge.

1. Where *should* the art student have drilled the hole? (Hint: Consider the right half of the circle $x^2 + y^2 = 1$, and see Solved Problem 8.2 for a discussion of computing a center of mass.)

2. Now that he has made the mistake, he decides that rather than drill a second hole, he will cut the piece with the hole in it in such a way that it will balance on the spike at the point of the hole. Explain clearly how the disk should be cut so that our artist friend can understand.

3. What is the area of the piece of the metal disk that balances at the point where the hole $(\frac{1}{2}, 0)$ was drilled?

Solved Problem 8.3: Arc length, area, and volume (CCH Text 8.2)

Let $f(x) = \dfrac{e^x + e^{-x}}{2}$ and $g(x) = e^x + e^{-x}$.

(a) Find the lengths of the curves, f and g, from $x = 0$ to $x = 1$, and compare the answers.

(b) Find the area bounded by the graph of f and the x-axis from $x = 0$ to $x = 1$.

(c) Find the volume of the solid generated if the region in Part (b) is rotated around the x-axis.

Solution to (a): We begin by defining f in Maple as an expression.

>y := (exp(x) + exp(-x))/2;

$$y := \frac{1}{2}e^x + \frac{1}{2}e^{(-x)}$$

Now arc length is the integral of $ds = \sqrt{1 + (dy/dx)^2}$ over an appropriate interval, so our next step is to compute dy/dx followed by ds.

>dy := diff(y, x);

$$dy := \frac{1}{2}e^x - \frac{1}{2}e^{(-x)}$$

>ds := sqrt(1 + dy^2);

$$ds := \sqrt{1 + \left(\frac{1}{2}e^x - \frac{1}{2}e^{(-x)}\right)^2}$$

To get the arc length, we use Maple's `int` command. Integrate ds over the interval $[0, 1]$ and then use `evalf` to get the decimal approximation.

>int(ds, x = 0..1);

$$\frac{1}{2}\frac{(e)^2 - 1}{e}$$

```
>evalf(");
```
$$1.175201194$$

If we redefine y as follows:

```
>y := exp(x) + exp(-x);
```
$$y := e^x + e^{(-x)}$$

---

then execute the above commands for dy, ds, and the integral again in an attempt to get the arc length for g. We see that Maple cannot perform the integration since it just returns the integral to us. (It is rare that an arc length integral can be evaluated exactly.)

```
>int(ds, x = 0..1);
```
$$\int_0^1 \sqrt{1 + \left(e^x - e^{(-x)}\right)^2}\, dx$$

However, using `evalf` will give the decimal approximation of 1.551889097.

The second function, g, is two times the first, f, so one might expect its arc length to be twice as much. As we see, that is not the case. Plot both f and g on the same axes to see why not.

Solution to (b): The area is $\int_0^1 e^x + e^{-x}\, dx$. (The reader should plot a picture of the region.) Since the integrand has already been defined at the end of Part (a), we need only to perform the integration. Using `evalf` directly on the output gives us the decimal approximation.

```
>evalf(int(y, x = 0..1));
```
$$2.350402387$$

Solution to (c): We continue assuming the definition $y = e^x + e^{-x}$ remains in effect. Slice the solid into "disks" with their centers on the x-axis. Each disk has volume $\pi y^2 \Delta x$. Thus the total volume is approximated by a Riemann sum of the form $\sum \pi y^2 \Delta x$. Therefore the volume is given by the integral $\int_0^1 \pi y^2\, dx$. We again use `evalf` directly on this integral to get a decimal approximation of the volume.

```
>evalf(int(Pi*y^2, x = 0..1));
```
$$17.67730332$$

Laboratory Exercise 8.3

Arc Length and Volume (CCH Text 8.2)

Name _____   Due Date _____

In each of the following, find the length of the given arc and the volume of the solid obtained by rotating the given region, R, about the x-axis.

1. Arc: $y = \sin x$ from $x = 0$ to $x = \pi$.

   R: One arch of the sine curve above the x-axis.

2. Arc: $y = x^2$ from $x = 0$ to $x = 1$.

   R: The region between $y = x^2$ and $y = \sin x$ in the first quadrant.

3. Arc: $y = \sqrt{1 - \frac{x^2}{9}}$ from $x = 0$ to $x = 3$. (The integral is a special case of an *elliptic integral*.)

   R: The region in the first quadrant that lies beneath $y = \sqrt{1 - \frac{x^2}{9}}$ and outside the unit circle.

Laboratory Exercise 8.4

Arc Length and Limits (CCH Text 8.2)

Name _____   Due Date _____

In this problem we will examine the length of the arc of the curve $y = x^n$ on the interval $[0, 1]$ for different values of n.

1. Approximate the length of the arc of the curve $y = x^n$ on the interval $[0, 1]$ for $n = 1, 10, 20,$ and 100.

2. For the case $n = 1$ explain how you can get the answer very quickly by just looking at the graph.

3. Discuss any pattern or trend you see in the calculations in Part 1.

4. Plot the graphs of the four curves in Part 1 and use them to help explain what is happening to the arc lengths as n gets larger.

5. Based on all the above, find $\lim_{n \to \infty} \int_0^1 \sqrt{1 + (n+1)^2 x^{2n}}\, dx$.

Hint: $\sqrt{1 + (n+1)^2 x^{2n}} = \sqrt{1 + \left(\dfrac{d}{dx} x^{n+1}\right)^2}$.

6. Repeat Parts 1 through 4 using the curve $y = \sqrt{1 - x^{2n}}$ on the interval $[-1, 1]$.

Laboratory Exercise 8.5

Ratio of Arc Length to Area (CCH Text 8.2)

Name _____   Due Date _____

We will explore what happens to the ratio of arc length to area on $[0, 1]$ as $a \to \infty$ for four curves that depend on the parameter a. For each of the four functions that follow,

(a) Plot the graph of the function for $a = 1$.

(b) Find the area bounded by the function and the x-axis on $[0, 1]$.

(c) With pencil and paper, write down the integral formulas for the arc length on $[0, 1]$ and the area under the curve on $[0, 1]$. Use these to find an integral formula for the limit of the ratio of arc length to area as $a \to \infty$. (<u>Hint</u>: Factor out a from each integral before taking the limit.)

(d) Using your work in Part (c), find the limit as $a \to \infty$ of the ratio of arc length to area on $[0, 1]$.

(e) By looking at the geometry of the graph, can you find a way to predict the limit in Part (d) without doing the calculations?

1. $a(x - x^2)$.

2. $a(\frac{1}{2} - |x - \frac{1}{2}|)$.

3. $a\sin(\pi x)$.

4. a times a semicircle of radius 1. (You figure out the function.)

Solved Problem 8.4: From the Earth to the moon (CCH Text 8.3)

(a) Find the work required to move a rocket ship of mass m from the surface of the Earth to the surface of the moon.

(b) What initial velocity must be given to a cannonball fired from the surface of the Earth if it is to land on the moon?

Solution to (a): In our solution we are going to take into account the influence of the moon's gravity. The following data will be necessary.

$G =$	Universal gravitational constant	$= 6.67 \times 10^{-11}$
$M =$	Mass of the Earth	$= 5.98 \times 10^{24}$ kilograms
$P =$	Radius of the Earth	$= 6.38 \times 10^{6}$ meters
$L =$	Mass of the moon	$= 7.35 \times 10^{22}$ kilograms
$Q =$	Radius of the moon	$= 1.74 \times 10^{6}$ meters
$C =$	Distance from center of Earth to center of moon	$= 3.84 \times 10^{8}$ meters

You can define each of the above constants in Maple on one line. Here is a partial entry showing you how. Notice that each definition is separated by a semicolon.

```
>G := 6.67*10^(-11); M := 5.98*10^24; P := 6.38*10^6; ...;
```

If r is the distance from the center of the Earth to the rocket ship, then the force pulling toward the Earth is $\dfrac{GMm}{r^2}$ while the force pulling it toward the moon is given by $\dfrac{GLm}{(C-r)^2}$. Thus the net force on the rocket is the difference between these two forces,

$$\text{Force} = \frac{GMm}{r^2} - \frac{GLm}{(C-r)^2}.$$

To obtain the work, we integrate the force from the surface of the Earth, $r = P$, to the surface of the moon, $r = C - Q$. If we define the variable Force as given above as an expression, then we can ask Maple to integrate it for us:

```
>int(Force, r = P..(C - Q));
```
$$.5867022268 \, 10^8 m$$

Thus the work done to move a rocket ship of mass m from the surface of the Earth to the surface of the moon will be $5.86702 \times 10^7 m$ joules. (Recall that a joule is a Newton-meter of work.)

Solution to (b): The cannonball of mass m must have sufficient kinetic energy, $\frac{1}{2}mv^2$, to pass the point where the gravitational attraction of the moon and the Earth are equal. From that point it will fall to the surface of the moon. To find that point, we need to solve the equation Force $= 0$ for r where Force is the variable defined above.

>solve(Force = 0, r);

$$.4318802515 \ 10^9, \quad .3456766773 \ 10^9$$

The first answer cannot be correct since it is greater than C, the distance from the Earth to the moon. We conclude that the equilibrium point must occur at $r = 3.45677 \times 10^8$ meters. The work required to lift the cannonball to this point is therefore

$$\int_P^{3.45677 \times 10^8} \frac{GMm}{r^2} - \frac{GLm}{(C-r)^2} \, dr.$$

Since the integrand is still the Force variable, we have

>int(Force, r = P..3.45677*10^8);

$$.6124937076 \ 10^8 m$$

---

so the work is $6.12494 \times 10^7 m$ joules.

We get the required velocity by setting the kinetic energy equal to this work and solving for v:

>solve(m*v^2/2 = .6124937076*10^8*m, v);

$$11067.91496, \quad -11067.91496$$

Only the result 11067.91496 meters per second makes sense since the other solution is negative.

Laboratory Exercise 8.6

From the Earth to the Sun (CCH Text 8.3)

Name _____   Due Date _____

1. Taking into account the gravitational pull of the sun, how much work is required to send a rocket ship of mass m from the surface of the Earth to the surface of the sun? You will need some of the data in Solved Problem 8.4 as well as the following data about the sun.

$$\begin{aligned} \text{Mass of sun} &= 1.97 \times 10^{30} \text{ kilograms} \\ \text{Radius of sun} &= 6.95 \times 10^{8} \text{ meters} \\ \text{Distance from Earth to sun} &= 1.49 \times 10^{11} \text{ meters} \end{aligned}$$

2. Explain what the sign of the answer means.

3. Find the point between the Earth and the sun where the pull of the Earth's gravity matches that of the sun.

4. Calculate the initial velocity of a cannonball fired from the surface of the Earth if it is to land on the surface of the sun.

5. Calculate the escape velocity for the sun.

Solved Problem 8.5: Rainfall in Anchorage (CCH Text 8.6)

Example 4 in Section 8.6 of the CCH Text uses the following normal probability distribution function with mean 15 and standard deviation 1 to describe the yearly rainfall in Anchorage, Alaska.

$$p(t) = \frac{1}{\sqrt{2\pi}} e^{-\frac{(t-15)^2}{2}}$$

(a) Find the probability that the rainfall in a given year is between 13 and 16 inches.

(b) Plot the graph of the cumulative distribution function.

(c) Find the smallest positive number k so that you can be 98% sure that the rainfall next year in Anchorage will be between $15 - k$ and $15 + k$ inches.

Solution to (a): We begin by defining the function p in Maple.

```
>p := t -> (1/sqrt(2*Pi))*exp(-(t - 15)^2/2);
```

$$p := t \to \frac{e^{(-1/2(t-15)^2)}}{sqrt(2\pi)}$$

Now we integrate $p(t)$ from 13 to 16.

```
>int(p(t), t = 13..16);
```

$$\frac{1}{2}\text{erf}\left(\frac{1}{2}\sqrt{2}\right) + \frac{1}{2}\text{erf}\left(\sqrt{2}\right)$$

$\text{erf}(x)$ is called the *error function* and is a function known to Maple. This function is defined as follows.

$$\text{erf}(x) = \frac{2}{\sqrt{\pi}} \int_0^x e^{-t^2} dt.$$

If we use **evalf**, we get the decimal approximation of the result as .8185946140. So the probability is approximately 0.82 that the rainfall will be between 13 and 16 inches.

Solution to (b): The cumulative distribution function is $P(x) = \int_{-\infty}^{x} p(t)\, dt$. We define this function in Maple with

```
>P := x -> int(p(t), t = -infinity..x);
```

$$P := x \to \int_{-\infty}^{x} p(t)\, dt$$

If we type P(x); at a Maple prompt, Maple will evaluate this in terms of erf(x). Try it. The result of the following command is the graph of the cumulative distribution function in Figure 8.3.

```
>plot(P(x), x = 0..40);
```

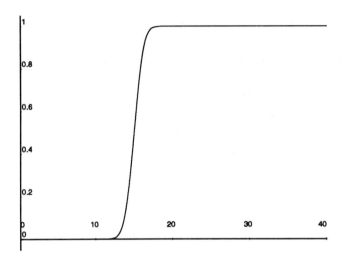

Figure 8.3: Cumulative distribution function

Solution to (c): The probability that the rainfall is within k inches of 15 is $\int_{15-k}^{15+k} p(t)\,dt$, so we ask Maple to evaluate this integral for us and call it prob.

```
>prob := int(p(t), t = 15 - k..15 + k);
```

$$prob := \operatorname{erf}\left(\frac{1}{2}\sqrt{2}k\right)$$

We want this probability to be 0.98, so we ask Maple to solve the equation $prob = 0.98$ for k. We use fsolve here.

```
>fsolve(prob = 0.98, k);
```
$$2.326347875$$

Thus, with 98% certainty we can predict next year's rainfall in Anchorage will be between 12.67 and 17.32 inches.

You may have noticed that there is a difficulty with the basic assumption that the rainfall is normally distributed. This means that we are allowing the possibility that the rainfall next year may be negative! According to the distribution function, theoretically this is possible. You should, however, evaluate $\int_{-\infty}^{0} p(t)\,dt$ to determine the probability that it will happen.

Remark: The student is encouraged to plot $p(t)$ and duplicate the graph given in the CCH Text.

Laboratory Exercise 8.7

SAT Scores (CCH Text 8.6)

Name _____   Due Date _____

According to the book, *For All Practical Purposes,* W.H. Freeman and Co., edited by Lynn Steen, SAT scores are normally distributed with a mean of 500 and a standard deviation of 100.

1. Write the probability function for SAT scores.

2. What percentage of students score between 500 and 550 on the SAT exam?

3. What percentage of students score more than 700 on the SAT exam?

4. Plot the graph of the cumulative distribution function.

5. Find a value of k so that you are 90% certain that a student selected at random will have an SAT score between $500 - k$ and $500 + k$.

6. The assumption of a normal distribution for SAT scores allows the possibility of negative scores on the exam. What percentage of students does this assumption predict will score less than zero? How serious do you think this flaw in the model is?

Chapter 9
Differential Equations

In this chapter we will use Maple as an aid in the qualitative study of differential equations. Although Maple can solve many differential equations in closed form (that is, exactly), this feature will not be emphasized here.

Solved Problem 9.1: Families of solutions (CCH Text 9.1)

Show that $e^{-x} + ce^{-2x}$ is a family of solutions of $y' + 2y = e^{-x}$. Plot this family of curves.

Solution: The first part of this problem can be solved with pencil and paper, but this will show how to use Maple. The first step is to write the equation in the form $y' + 2y - e^{-x} = 0$. Now we define the expression on the left-hand side in Maple, keeping in mind that y is a function of x.

>de := diff(y(x), x) + 2*y(x) - exp(-x);

$$de := \left(\frac{\partial}{\partial x} y(x)\right) + 2y(x) - e^{(-x)}$$

Next, define the family as a function.

>f := x -> exp(-x) + c*exp(-2*x);

$$f := x \to e^{(-x)} + ce^{(-2x)}$$

Now we substitute the function f for the function y in the differential equation de and then simplify:

>subs(y=f, de);

$$de := \left(\frac{\partial}{\partial x} f(x)\right) + 2f(x) - e^{(-x)}$$

>simplify(");

$$0$$

The fact that the result of this substitution is 0 shows that we have a solution of the equation.

To obtain a graph of the solutions, we first define a set containing the functions in the family for values of c ranging from -3 to 3 in steps of 0.2.

```
>family := {seq(subs(c = i/5, f(x)), i = -15..15)};
```

As a result of this command, Maple will list the expressions comprising the set `family`. All we need to do now is to plot this set. The result of the following command appears in Figure 9.1.

```
>plot(family, x = -3..1, y = -10..10);
```

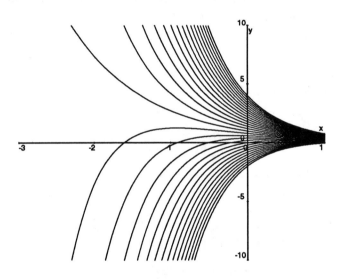

Figure 9.1: Solutions of $y' + 2y = e^{-x}$

Laboratory Exercise 9.1

Families of Solutions (CCH Text 9.1)

Name _____   Due Date _____

1. Show that $e^{-x}(A\sin x + B\cos x)$ is a family of solutions of $y'' + 2y' + 2y = 0$. (Enter y'' as `diff(y(x), x$2)`.)

2. Graph 10 members of each of the two families $Ae^{-x}\sin x$ and $Be^{-x}\cos x$.

3. For what value of k is $e^{-\frac{x}{2}}(A\sin(kx) + B\cos(kx))$ a solution of $y'' + y' + y = 0$? (Hint: Substitute $e^{-\frac{x}{2}}(A\sin(kx) + B\cos(kx))$ into the differential equation and see what this tells you about k.)

Solved Problem 9.2: Slope fields (CCH Text 9.2)

Sketch the slope field for the differential equation $2xy + (1+x^2)y' = 0$. Include several solution curves in the picture.

Solution: For this problem we will use a function in another of Maple's special packages, DEtools, so we will begin by loading this package into Maple's memory.

>with(DEtools);

When this command is executed, Maple will print a list of the functions in this package that are now available to us. The one we will use is DEplot1. Its purpose is to plot slope fields for first-order differential equations. In addition, it will also plot a solution curve on the slope field. The syntax for DEplot1 is as follows:

DEplot1(expr, varlist, xrange, yrange)

Assuming your differential equation is in the form $y' = f(x,y)$, then expr is $f(x,y)$. varlist is a list of the variables that appear in expr. xrange is a plot range of the form x = a..b. yrange is also a plot range of the form y = c..d, but this argument may be replaced by a set of initial values whose solution curves will be found and plotted on the slope field.

Our differential equation may be written in the form $y' = -\dfrac{2xy}{1+x^2}$; therefore the first argument to DEplot1 is the right-hand side of the equation. The remaining arguments are given in the following command whose result appears in Figure 9.2a.

>DEplot1(-2*x*y/(1 + x^2), [x, y], x = -2..2, y = -2..2);

To produce a slope field with several solution curves, we need to replace that last argument in DEplot1 with a set of initial values for the solution curves. We will plot the solution curves for $y = -2, -1.5, -1, -0.5, 0, 0.5, 1, 1.5$, and 2 and $x = 0$ for each of these values. The initial conditions are specified as points through which the solution curve must pass. These points may be produced in Maple as follows:

>inits := {seq([0, 0.5*i], i = -4..4)};

$inits := \{[0, -.5], [0, .5], [0, -2.0], [0, -1.5], [0, 0], [0, -1.0], [0, 1.0], [0, 1.5], [0, 2.0]\}$

Now use the set *inits* as the last argument to DEplot1 in place of the yrange argument. The result appears in Figure 9.2b.

```
>DEplot1(-2*x*y/(1 + x^2), [x, y], x = -2..2, inits);
```

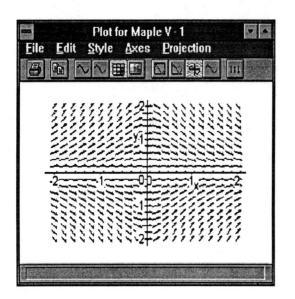 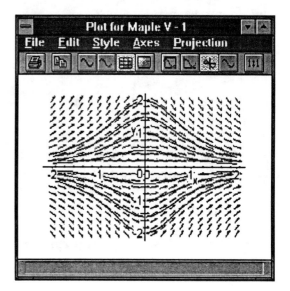

a. b.

Figure 9.2: Slope field for $2xy + (1 + x^2)y' = 0$

<u>Remark:</u> Once your slope field is produced as in Figure 9.2a, you can sketch in several solutions by hand. Your hand-drawn curves should be similar to those displayed in Figure 9.2b.

Laboratory Exercise 9.2

Slope Fields (CCH Text 9.2)

Name _____   Due Date _____

For each of the following equations, use Maple to make two graphs. The first graph is to be the slope field of the equation, and the second is to be the slope field together with several solution curves plotted on it. You should check that your second graph makes sense by sketching several solution curves by hand on your first graph.

1. $(x^2 + y^2)y' = xy$

2. $y' = \sin(xy)$

3. $y' = x^2 - y^2$

Solved Problem 9.3: Euler's method (CCH Text 9.3)

Let $y(x)$ be a solution of $y' = 1 + 2xy$ that passes through the point $(0, -1)$.

(a) Use 5 steps of Euler's method to approximate the graph of $y(x)$ on the interval $x = 0$ to $x = 1$.

(b) Use 20 steps of Euler's method to approximate the graph of $y(x)$ on the interval $x = 0$ to $x = 1$.

(c) Use 20 steps of Euler's method to approximate the value of $y(1)$.

Solution to (a): We will use the procedure EULER that appears in Appendix II. Be sure that you have entered and saved the file as instructed. Now read the procedure into Maple's memory with

```
>read 'EULER';
```

The syntax for EULER is as follows:

$$\text{EULER(expr, varlist, inivalue, h, n)}$$

Assuming your differential equation is in the form $y' = f(x, y)$, then expr is just $f(x, y)$. The second argument, varlist, is a list of the variables appearing in expr. The third argument, inivalue, is the starting point or *initial value*. The third and fourth arguments are the stepsize Δx and the number of steps to be executed.

We want to execute 5 steps of Euler's method beginning at $x = 0$ and ending at $x = 1$ so we choose $\Delta x = (1 - 0)/5 = 0.2$.

```
>EULER(1 + 2*x*y, [x, y], [0,-1], 0.2, 5);
```

$$\begin{array}{rl} 0, & -1. \\ .2, & -.8 \\ .4, & -.664 \\ .6, & -.57024 \\ .8, & -.5070976 \\ 1.0, & -.469368832 \end{array}$$

EULER thus produces a table of approximate values for y. In order to make it easy to plot these points, EULER also produces a set, called `pointset`, which essentially contains these points. pointset may be plotted directly. See Figure 9.3a for the result of the following command.

```
>plot(pointset, x = 0..1, y = -1..0);
```

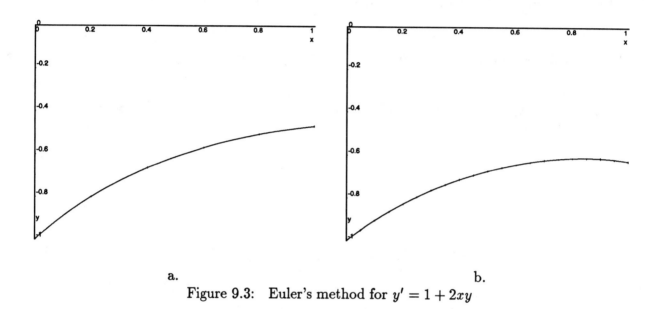

a. b.
Figure 9.3: Euler's method for $y' = 1 + 2xy$

Solution to (b): To move from 0 to 1 in 20 steps, we choose $\Delta x = (1-0)/20 = 0.05$. We do not reproduce here the result of the next command, but the graph of the following plot command can be seen in Figure 9.3b.

```
>Euler(1 + 2*x*y, [x, y], [0,-1], 0.05, 20);
```

```
>plot(pointset, x = 0..1, y = -1..0);
```

Solution to (c): The last entry in the output of the previous EULER command is

$$1.00, -.6235436348.$$

Thus the value of $y(1)$ is $-.6235436348$. This approximation can be improved upon by using still smaller step sizes.

Laboratory Exercise 9.3

Estimating Function Values with Euler's Method (CCH Text 9.3)

Name _____   Due Date _____

It is known that the solution curve of the differential equation $y' = \dfrac{y \cos x}{1 + 2y^2}$ that passes through $(0, 1)$ and satisfies $y(1) = 1.27494$, $y(2) = 1.28718$, and $y(10) = 0.813712$.

1. Execute Euler's method starting at $(0, 1)$ using step size $\Delta x = 0.5$ and $n = 20$. Graph the approximate solution you obtain.

2. What estimates did you get in Part 1 for $y(1)$, $y(2)$, and $y(10)$? Compare these answers with the given values, and discuss how the accuracy of Euler's method varies with the distance from the starting point.

3. Repeat Part 1 using step size $\Delta x = 0.1$ and $n = 100$.

4. What estimates did you get in Part 3 for $y(1)$, $y(2)$, and $y(10)$? Compare these estimates with those you obtained in Part 1 and with the given values. Discuss how the accuracy of Euler's method varies with the step size.

> **Solved Problem 9.4: A tank of water (CCH Text 9.6)**

A water tank initially contains V_0 cubic feet of water. Water runs into the tank at a constant rate of 36 cubic feet per minute, but a valve allows water to flow out of the tank at a rate equal to 20% of the volume of water in the tank per minute.

(a) Set up and solve a differential equation that determines the volume of water in the tank as a function of time. Express your solution in terms of t and V_0.

(b) Graph the solutions for $V_0 = 100, 200, \ldots, 1000$.

(c) Find the equilibrium solution. Add the graph of the equilibrium solution to your graph in Part (b) and determine if it is stable or unstable.

Solution to (a): The rate of change of V is the rate the water is coming in minus the rate the water is going out. Therefore

$$\frac{dV}{dt} = \text{rate in} - \text{rate out} = 36 - 0.2V.$$

The solution of this differential equation is $V = 180 - ke^{-0.2t}$ (which you should be able to find by hand). When $t = 0$, the equation is $V_0 = 180 - k$. Thus $k = 180 - V_0$, and the solution in terms of V_0 is $V = 180 - (180 - V_0)e^{-0.2t}$.

Solution to (b): We will use Maple's `seq` command to generate the set of functions for $V_0 = 100, 200, \ldots, 1000$. We will do this by computing $V_0 = 100 * i$ for $i = 1, \ldots, 10$.

```
>SET := {seq(180 - (180 - 100*i)*exp(-0.2*t), i = 1..10)};
```

(Do *not* use lower case letters for the word "set.") As a result of the above command, Maple will list the 10 elements of SET after making some simplifications to the functions themselves. Now we ask Maple to plot SET. The graph appears in Figure 9.4a.

```
>plot(SET, t = 0..10);
```

Solution to (c): The equilibrium solution occurs when $\frac{dV}{dt} = 0$, that is, when $V = 180$. (Note that we obtained this without resorting to the solution of the differential equation.) The horizontal line $V = 180$ may be added to SET with the following command; then SET may be replotted using the previous `plot` command. See Figure 9.4b in which the second graph from the bottom is the graph of $V = 180$.

```
>SET := SET union {180};
```

From the graphs we see that the solutions both above and below bend toward the equilibrium solution. We conclude that the equilibrium solution is stable. We can also verify this analytically by noting that for any V_0, $\lim_{t \to \infty} 180 - (180 - V_0)e^{-0.2t} = 180$.

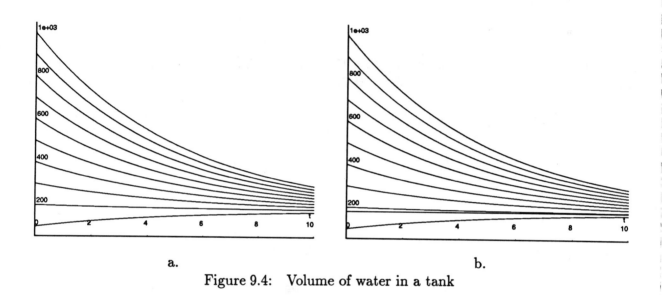

a. b.

Figure 9.4: Volume of water in a tank

Laboratory Exercise 9.4

A Leaky Balloon (CCH Text 9.6)

Name _____   Due Date _____

A leaky balloon initially contains V_0 cubic inches of air. Additional air is being blown into the balloon at a constant rate of 24 cubic inches per minute. The balloon leaks air through a small hole at a rate equal to 3% of its volume at any given time.

1. Set up and solve a differential equation that gives the volume of air in the balloon as a function of time t. Express your answer in terms of t and V_0.

2. Find the equilibrium solution and explain what it means in practical terms.

3. Graph the equilibrium solution as well as several other nearby solution curves.

4. Is the equilibrium solution stable? Justify your answer both analytically and graphically.

5. What is the diameter of the balloon when the equilibrium solution is attained?

Laboratory Exercise 9.5

Baking Potatoes (CCH Text 9.5)

Name _____ Due Date _____

A chef realizes at 6:30 that he has forgotten to preheat the oven for baked potatoes to be served with dinner at 7:30. He turns on the oven and puts in the potatoes. The oven temperature t minutes after it is turned on is $O(t) = 400 - 325e^{-0.7t}$ degrees Fahrenheit. The potatoes heat according to Newton's law: The rate of change in the temperature, $P(t)$, of the potatoes is $\frac{dP}{dt} = 0.03(O(t) - P(t))$. The potatoes will be done when they reach a temperature of 270 degrees.

1. Set up a differential equation whose solution gives the temperature of the potatoes at time t. (Do not try to solve it.)

2. What is the temperature of the oven when $t = 0$? (You may assume this is also the initial temperature of the potatoes.)

3. Use Euler's method to approximate and graph a solution of the equation. We recommend a step size of 0.2. Get your starting point from Part 2 and choose n large enough to find out when the potatoes are done. (If they aren't done in an hour, the chef will probably be fired.)

4. According to your solution, when will the potatoes be done?

5. What time would the potatoes have been done if the chef had remembered to preheat the oven to 400 degrees? (You should be able to set up and solve a differential equation to answer this.)

6. **(Optional):** Perform an experiment at home to determine if the times predicted in this exercise are reasonable.

Solved Problem 9.5: Terminal velocity (CCH Text 9.6)

The velocity (in feet per second) of a falling object subject to air resistance is governed by the differential equation $\frac{dv}{dt} = 32 - 0.4v^{\frac{3}{2}}$.

(a) What is the terminal velocity of the object?

(b) Use Euler's method to produce a graph of the velocity as a function of time if the initial velocity is zero.

(c) How long does it take for the object to reach 99% of its terminal velocity?

Solution to (a): Terminal velocity occurs when the rate of change in velocity is zero. That is, when $32 - 0.4v^{\frac{3}{2}} = 0$. Solving this equation gives $v = 18.56635533$.

Solution to (b): We know from experience that the object will approach its terminal velocity quickly, so we use Euler's method on the interval $t = 0$ to $t = 3$. After loading EULER into Maple's memory with **read 'EULER'**; we execute

```
>EULER(32 - 0.4*v^(3/2), [t, v], [0, 0], 0.01, 300);
```

As a result of this command, Maple will produce a rather long list of points, 300 to be exact. We can plot these points by recalling that EULER also produces a set called **pointset** which can be plotted. Before doing so, we add the horizontal line $v = 18.56635533$ to the graph with

```
>pointset := pointset union {18.56635533};
```

The result of the following command appears in Figure 9.5:

```
>plot(pointset, x = -1..10);
```

Solution to (c): 99% of the terminal velocity found in Part (a) is, to three decimal places, 18.381. We now search back through the output of the EULER command until we find a second coordinate near this number. We find that 99% of the terminal velocity is reached at approximately 1.89 seconds.

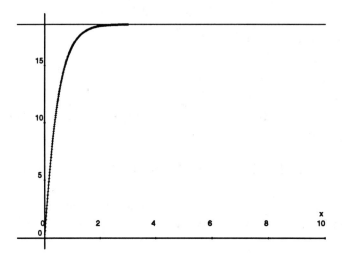

Figure 9.5: Estimating time to terminal velocity

Laboratory Exercise 9.6

Drag and Terminal Velocity (CCH Text 9.6)

Name _____   Due Date _____

This is a continuation of Solved Problem 9.5.

1. The coefficient, 0.4, of $v^{3/2}$ in Solved Problem 9.5 represents *drag*. Use Euler's method to produce the graph of velocity against time if the drag is doubled and if it is tripled. Add both to the graph from Solved Problem 9.5.

2. How long does it take to reach 99% of the terminal velocity if the drag is doubled? How long if the drag is tripled?

3. A feather and a cannonball are dropped at the same time. Which reaches 99% of its terminal velocity first? Explain.

Solved Problem 9.6: Population growth with a threshold (CCH Text 9.7)

A population, $P(t)$ at time t, is governed by the following differential equation.
$$\frac{dP}{dt} = -P\left(1 - \frac{P}{500}\right)\left(1 - \frac{P}{2000}\right)$$

(a) Find the equilibrium solutions. Plot $\frac{dP}{dt}$ as a function of P and use the graph to classify each equilibrium solution as stable or unstable.

(b) Use Euler's method to obtain approximate solutions for initial populations $P(0) = 200, 400, 600, 1000, 2500$, and 3000. Graph each of these solutions together with the equilibrium solutions found in Part (a).

(c) Explain in practical terms what these graphs say about the population in question.

Solution to (a): Equilibrium solutions occur where $\frac{dP}{dt} = 0$. That is, when $-P(1 - \frac{P}{500})(1 - \frac{P}{2000}) = 0$, or when P is 0, 500 or 2000.

To get the graph of $\frac{dP}{dt}$ as a function of P, we execute the command given below. The resulting graph appears in Figure 9.6. (It is worth repeating the warning in the CCH Text that this is *not* the graph of P as a function of t; rather, it is the graph of the rate of change in population as a function of population.)

```
>plot(-P*(1 - P/500)*(1 - P/2000), P = -100..2500, y = -500..800);
```

From the graph in Figure 9.6, we see that when $P < 500$, $\frac{dP}{dt}$ is negative so that P is a decreasing function of t. Thus if $P < 500$, the population decreases toward 0 and we conclude that $P = 0$ is a stable equilibrium solution.

On the interval $500 < P < 2000$, $\frac{dP}{dt}$ is positive, and thus P is increasing on this interval. But P is decreasing for $P > 2000$. From this analysis, populations between 500 and 2000 will increase toward 2000, and populations over 2000 will decrease toward 2000. Thus $P = 500$ is an unstable equilibrium solution, and $P = 2000$ is a stable equilibrium solution.

Solution to (b): Read EULER into Maple's memory with

```
>read 'EULER';
```

We will use $\Delta x = 0.05$ with $n = 100$ steps.

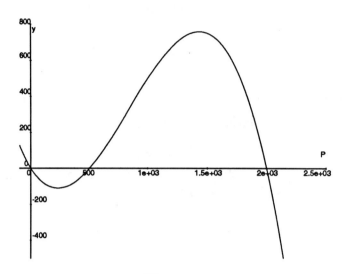

Figure 9.6: $\dfrac{dP}{dt}$ as a function of P

We will also use a sixth argument to EULER. This sixth argument can be anything, but we will use 1. Its purpose is to suppress the output but still allow us to have access to **pointset**. This makes it easy for us to use EULER to plot the graphs for this part. To do so we need to create a set with the objects we wish to plot. We start with

>SET := {500, 2000};
$$SET := \{500, 2000\}$$

Next, we create a list with the values of $P(0)$ for the initial conditions.

>L := [200, 400, 600, 1000, 2500, 3000];
$$L := [200, 400, 600, 1000, 2500, 3000]$$

Now enter the following exactly as it appears here. Pay close attention to the syntax; in particular, note that the set assignment ends with a colon. This is important. Failure to do this will result in a very long output list.

>for j in L do EULER(-P*(1 - P/500)*(1 - P/2000), [t,P],[0,j], .05,100,1):
SET := SET union pointset: od:

Depending on your computer, this command may take some time to execute. There will be no visible output. When Maple is finished executing this command, it will produce the separator line. This is your indication that Maple is done.

SET now contains the items we wish to plot. The result of the following command appears in Figure 9.7:

>plot(SET, x = 0..4);

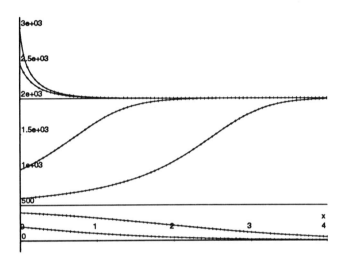

Figure 9.7: Equilibrium and other solutions

These are plots of P as functions of time. We leave it as an exercise for the reader to label each graph with the starting value of P that produced it. Observe that solution curves above and below $P = 2000$ tend toward that solution verifying that it is stable, while curves near $P = 500$ move away indicating that this solution is unstable. This concurs with our earlier classifications of equilibrium solutions.

Solution to (c): This is an example of a population model with a *threshold*. If the population ever drops below 500 individuals it will eventually die out. (Perhaps there is an insufficient number of mating pairs to sustain the population.)

If the population is above the threshold of 500 but below 2000, it can be expected to increase toward 2000, but populations above 2000 quickly decrease to that value. This indicates that a population level of 2000 may be the largest that the environment can support and that if we start with a population of more than 500 individuals we can expect to see the population eventually stabilize near 2000.

Laboratory Exercise 9.7

Comparing Population Models (CCH Text 9.7)

Name _____   Due Date _____

The Gompertz equation, $\frac{dP}{dt} = 0.7P \ln\left(\frac{2000}{P}\right)$, is used for population modeling.

1. Graph $\frac{dP}{dt}$ as a function of P. Find the equilibrium points and determine which are stable and which are unstable.

2. Use Euler's method to approximate 10 solutions of the Gompertz equation. Graph these solutions with all equilibrium solutions.

3. Compare the Gompertz model with the logistic model discussed in Solved Problem 9.6. Your discussion should include equilibrium solutions, thresholds, carrying capacity, and rates of population growth or decline.

Solved Problem 9.7: Spider mites and lady bugs (CCH Text 9.8)

Spider mites infest a gardener's plot of marigolds. Lady bugs eat spider mites. The number of spider mites, S, and lady bugs, L, are governed by the following system of differential equations.

$$\frac{dS}{dt} = 2S - SL$$
$$\frac{dL}{dt} = -6L + 3SL$$

(a) Find the equilibrium points.

(b) Plot the slope field for the phase plane near the equilibrium points and include several trajectories in your picture.

(c) Discuss how the two populations can be expected to vary with time.

Solution to (a): The equilibrium points occur where both $2S - SL = 0$ and $-6L + 3SL = 0$. This system of equations is solved easily to obtain the equilibrium points $(0,0)$ and $(2,2)$.

Solution to (b): In order to obtain the slope field for the phase plane, we will use another routine from the DEtools package. We read the DEtools package into Maple's memory by executing

>with(DEtools);

The routine we need here is DEplot2. We will use two forms.

DEplot2([de1, de2], varlist, trange, xrange, yrange)

DEplot2([de1, de2], varlist, trange, inits)

The first form is used when we only want the slope field. The first argument is a list of differential equations, but since the derivatives are already isolated, we need only list the right-hand sides. Thus the slope field is plotted as follows. See Figure 9.8a.

>DEplot2([2*S - S*L, -6*L + 3*S*L], [S, L], t = 0..2 S = 1..3, L = 1..3);

To include trajectories in the plot, we use the second form given above. Replace the xrange and yrange arguments in DEplot2 with a set containing one or more initial points specified as [t, S, L]. The result of the following command is shown in Figure 9.8b. Note the direction of the arrows on the slopes. They indicate that the trajectories move in a counterclockwise fashion.

```
>DEplot2([2*S - S*L, -6*L + 3*S*L], [S, L], t = 0..2, {[1, 2, 1],
[1, 2, 1.5], [1, 2, 1.75]});
```

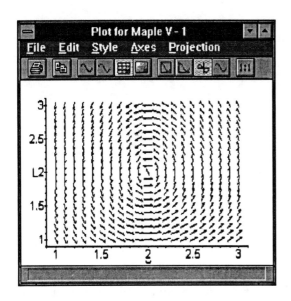
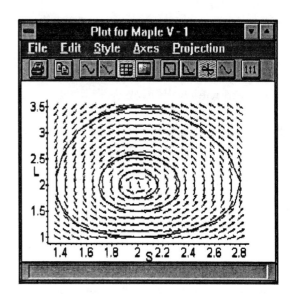

a.   b.

Figure 9.8: Spider mites and lady bugs

Solution to (c): The trajectories appear to be somewhat elliptical centered around the equilibrium point $(2, 2)$. As with the example in Section 9.8 of the CCH Text involving robins and worms, we expect the populations to vary cyclically. If we are at the bottom of one of the ellipses, there aren't enough spider mites (S-axis) to feed the lady bugs, so lady bugs die and the spider mite population increases. Moving counterclockwise, toward the left-hand end of the ellipse, we see that there are enough spider mites to reproduce rapidly and enough to support the diminished lady bug population. Thus both populations increase. At the top of the ellipse, there are so many spider mites that the lady bug population can indulge itself at the expense of the spider mites. Finally at the right-hand end of the ellipse, the spider mite population has dwindled to the point that the large population of lady bugs cannot get enough food. Both populations decrease until we return once more to the bottom.

Laboratory Exercise 9.8

Foxes and Hares (CCH Text 9.8)

Name _____   Due Date _____

Foxes prey on hares. The population of foxes, $F(t)$, and hares, $H(t)$, are governed by the following system of equations.

$$\frac{dF}{dt} = 3F - 2FH$$
$$\frac{dH}{dt} = -5H + 2FH$$

1. Find the equilibrium points.

2. Plot the slope field for the phase plane near the equilibrium points and add several trajectories to your picture.

3. Discuss how the two populations can be expected to vary with time.

Solved Problem 9.8: Springs and second-order equations (CCH Text 9.10)

The position of a moving spring at time t is given by $A\sin(\omega t + \phi)$. Investigate what happens as the constants A, ω, and ϕ vary. How do they relate to the physical characteristics of the spring and its motion?

Solution: We set $\omega = \phi = 1$ and examine the graphs as A varies. To get the graph in Figure 9.9a, we vary A from 0 to 2 in steps of 0.5. That is, we plot $A\sin(t+1)$ for $A = 0, 0.5, 1, 1.5, 2$. We first create a set with these functions as elements by letting $A = n/2$ for $n = 0, 1, 2, 3, 4$.

```
>R := {seq((n/2)*sin(t + 1), n = 0..4)};
```

$$R := \{\sin(t+1), \frac{1}{2}\sin(t+1), 2\sin(t+1), \frac{3}{2}\sin(t+1), 0\}$$

Now we plot R to get Figure 9.9a.

```
>plot(R, t = -2*Pi..2*Pi);
```

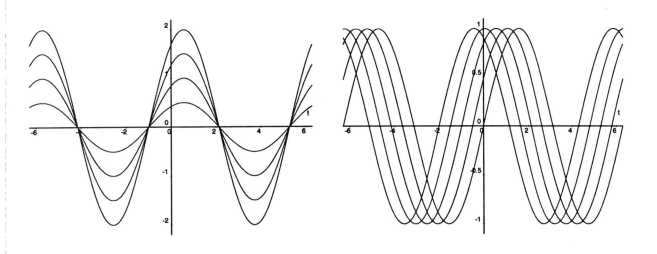

a.
b.
Figure 9.9: Varying amplitude and phase shift

Notice that as A increases, the amplitude of the curve increases, but nothing else changes. The larger values of A correspond to a larger maximum displacement of the spring.

To examine changes in the phase shift, ϕ, we put $A = \omega = 1$ and vary ϕ. The picture in Figure 9.9b shows these graphs for $\phi = 0, .5, 1, 1.5, 2$. We create the set as above with $\phi = n/2$ for $n = 0, 1, 2, 3, 4$ and then plot.

```
>S := {seq(sin(t + n/2), n = 0..4)};
```

$$S := \{\sin(t+1), \sin\left(t+\frac{3}{2}\right), \sin\left(t+\frac{1}{2}\right), \sin(t+2), \sin(t)\}$$

```
>plot(S, t = -2*Pi..2*Pi);
```

Notice that changing the phase shift changes the time when the maximum displacement of the spring occurs.

Lastly, we examine the effect of ω in the same way: $A = \phi = 1$ and ω varies from 0 to 2 in steps of 0.5. See Figure 9.10.

```
>T := {seq(sin((n/2)*t + 1), n = 0..4)};
```

$$T := \{\sin(t+1), \sin(2t+1), \sin\left(\frac{3}{2}t+1\right), \sin(1), \sin\left(\frac{1}{2}t+1\right)\}$$

```
>plot(T, t = -2*Pi..2*Pi);
```

Notice that as ω increases the period of oscillation decreases. Thus the parameter ω determines the rate of oscillation of the spring—larger values of ω correspond to a stronger spring.

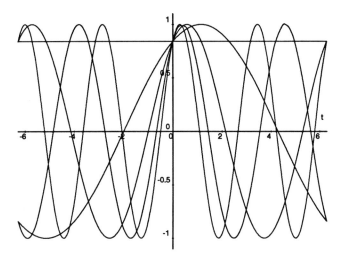

Figure 9.10: Varying the rate of oscillation

Laboratory Exercise 9.9

A Damped Spring(CCH Text 9.11)

Name _____   Due Date _____

In this laboratory, we will explore the movement of a spring in the underdamped, critically damped, and overdamped cases.

1. In the underdamped case, the movement of the spring is given by $e^{-at}(C_1 \cos Bt + C_2 \sin Bt)$. We will examine the special case, $C_1 = C_2 = 1$.

   (a) What is the initial displacement of the spring?
   (b) What is the initial velocity of the spring?
   (c) Produce graphs to show the effects of varying a and B.
   (d) Discuss the physical interpretation of a and B.

2. In the critically damped case, the movement of the spring is given by $(C_1 + C_2)e^{-at}$. Again, we specialize to the case $C_1 = C_2 = 1$.

   (a) What is the initial displacement of the spring?
   (b) What is the initial velocity of the spring?
   (c) Produce graphs to show the effects of varying a.
   (d) What is the physical interpretation of a?

3. In the overdamped case, the movement of the spring is given by $C_1 e^{At} + C_2 E^{Bt}$. We take $C_1 = C_2 = 1$.

   (a) What is the initial displacement of the spring?
   (b) What is the initial velocity of the spring?
   (c) Produce graphs to show the effects of varying A and B.
   (d) What are the physical interpretations of A and B?

Chapter 10
Approximations

In this chapter we will examine methods of approximating functions by polynomials and extend this idea to Taylor series. We will also look at Fourier series, which are expansions in terms of periodic functions.

Solved Problem 10.1: Approximating $\cos x$ (CCH Text 10.1)

(a) Find the tenth-degree Taylor polynomial, $P(x)$, for $\cos x$ about $x = 0$ and plot the two graphs on the same axes.

(b) Compare the values of $\cos x$ and $P(x)$ for $x = 1, 3$, and 5. Discuss how the size of x affects the accuracy of the approximations.

Solution to (a): Maple has a routine for computing Taylor series called, appropriately enough, `taylor`. The syntax for `taylor` is
$$\texttt{taylor(expr, var = a, n)}$$
where `expr` is the expression whose Taylor series we are seeking, `var = a` indicates the point about which we want to expand the series, `var` is the variable that appears in `expr`, and n is a positive integer specifying the order of the series. To find the tenth degree Taylor polynomial for $\cos x$, we first find its Taylor series giving it the name `taylorcos`.

```
>taylorcos := taylor(cos(x), x = 0, 11);
```

$$taylorcos := 1 - \frac{1}{2}x^2 + \frac{1}{24}x^4 - \frac{1}{720}x^6 + \frac{1}{40320}x^8 - \frac{1}{3628800}x^{10} + O(x^{11})$$

The $O(x^{11})$ at the end means that there are other terms in this series starting at powers of 11. Now we can see why we specified $n = 11$ in the `taylor` command: to display terms with powers up to 10 as required. We are still not done. We now *convert* this series to a polynomial using Maple's `convert` command and call this polynomial `polycos`. The second argument, `polynom`, tells Maple to convert the series to a polynomial.

```
>polycos := convert(taylorcos, polynom);
```

$$polycos := 1 - \frac{1}{2}x^2 + \frac{1}{24}x^4 - \frac{1}{720}x^6 + \frac{1}{40320}x^8 - \frac{1}{3628800}x^{10}$$

The graphs of both *polycos* and cos x appear in Figure 10.1a and are the result of the following command. We leave it to the reader to determine which one is which.

```
>plot({polycos, cos(x)}, x = -3*Pi..3*Pi, y = -2..2);
```

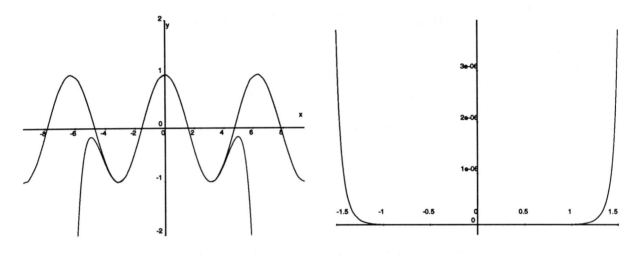

a.   b.

Figure 10.1: Taylor polynomial, cos x, and the error function

Solution to (b): One commonly used tool to judge the quality of approximations is the *relative error*. In our case, it is $\frac{\cos x - P(x)}{\cos x}$. This is useful because if we take the absolute value and multiply by 100, we get the error as a percentage of the size of the function. Let's first convert the Taylor polynomial into a function. We can do this with Maple's unapply command:

```
>P := unapply(polycos, x);
```

$$P := x \to 1 - \frac{1}{2}x^2 + \frac{1}{24}x^4 - \frac{1}{720}x^6 + \frac{1}{40320}x^8 - \frac{1}{3628800}x^{10}$$

We next define the relative error as a function of x with

```
>err := x -> (cos(x) - P(x))/cos(x);
```

$$err := x \to \frac{\cos(x) - P(x)}{\cos(x)}$$

We can see the values cos(1), $P(1)$, and the relative error if we execute

```
>evalf(cos(1)); evalf(P(1)); evalf(err(1));
```
$$.5403023059$$
$$.5403023038$$
$$.3886713007 \; 10^{-8}$$

So $P(1)$ is only off by 0.00000039%. We leave it to the reader to show that $P(3)$ misses by 0.10672914% but that $P(5)$ is off by more than 150%. It appears that $P(x)$ gives good approximations to $\cos x$ for x near the origin, but gives poor approximations for larger values of x. Figure 10.1b contains a graph of the relative error function.

Laboratory Exercise 10.1

Taylor Polynomials and the Cosine Function:
The Expansion Point (CCH Text 10.1)

Name _____   Due Date _____

In this exercise you will explore two strategies for approximating $\cos x$ for large values of x.

1. Let $P(x)$ be the eighteenth degree Taylor polynomial for the cosine function expanded about $x = 0$. Plot the graphs of $\cos x$ and $P(x)$ on the same axes.

2. Plot the graph of the relative error function and discuss the accuracy of using $P(x)$ as an approximation for $\cos x$ on the interval $[-\pi, \pi]$.

3. Since the cosine function has period 2π, $\cos(10) = \cos(10 - 4\pi)$. Show how to use this fact to obtain an accurate approximation of $\cos(10)$ using the Taylor polynomial from Part 1. Calculate the relative error of the approximation.

4. Use the Taylor polynomial from Part 1 to obtain an approximation for cos(2438762) and report the relative error. (Hint: Solve $2k\pi = 2438762$, choose an integer near the solution, and then follow the idea in Part 3.)

5. An alternative strategy for approximating cos(10) is to expand the cosine function about a point that is closer to 10.

   (a) Find the sixth-degree Taylor polynomial for $\cos x$ using expansion point 3π.

   (b) Use the Taylor polynomial you found in Part (a) to approximate cos(10) and report the relative error.

Laboratory Exercise 10.2

Taylor Polynomials and the Cosine Function: The Degree (CCH Text 10.2)

Name _____   Due Date _____

1. On the same axes plot $\cos x$ and its Taylor polynomials about $x = 0$ of degrees 0 through 10. You can create all the Taylor polynomials at once and place them in a set for plotting with

    ```
    {seq(convert(taylor(cos(x), x = 0, n), polynom), n = 0..10)};
    ```

2. Discuss the relationship between the cosine function and its Taylor polynomial of degree n as n increases.

3. Find (by trial and error) the Taylor polynomial about $x = 0$ of smallest degree for the cosine function that approximates $\cos(10)$ to three digits of accuracy.

4. On the same axes, plot $\ln x$ and its Taylor polynomials about $x = 1$ of degrees 0 through 10.

5. Discuss the relationship between the function $\ln x$ and its Taylor polynomial of degree n as n increases. Compare the picture here with the one in Part 1 and explain any similarities *and differences* you observe between the behavior of the Taylor polynomials in the two cases.

Solved Problem 10.2: Intervals of convergence (CCH Text 10.2)

Plot $\dfrac{1}{x^2+7}$ and several of its Taylor polynomials about $x = 0$. Use this to estimate the interval of convergence of its Taylor series.

Solution: First define $\dfrac{1}{x^2+7}$ in Maple as the variable z:

```
>z := 1/(x^2 + 7);
```
$$z := \frac{1}{x^2+7}$$

We now create a set s containing all the Taylor polynomials as we did in Lab Exercise 10.2.

```
>s := {seq(convert(taylor(z, x = 0, n), polynom), n = 1..10)};
```

As a result of this command, Maple will print out the set s containing the Taylor polynomials of degrees 1 through 10. Next, include z in s with

```
>s := s union {z};
```

Lastly, we plot the set s. See Figure 10.2.

```
>plot(s, x = -4..4, y = -0.2..0.4);
```

We leave it to the reader to determine which graph in Figure 10.2 is the graph of $\dfrac{1}{x^2+7}$. From this picture, the Taylor polynomials appear to stay close to $\dfrac{1}{x^2+7}$ from about $x = -2.5$ to $x = 2.5$. Outside this interval, the separation in the graphs is apparent. Thus the interval of convergence is approximately $(-2.5, 2.5)$. We emphasize that it is not possible to determine this interval exactly by looking at the graphs; we can only estimate it.

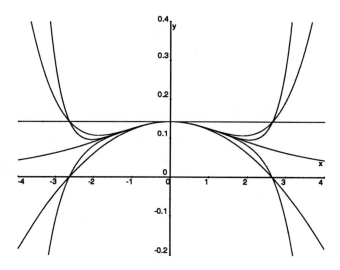

Figure 10.2: Taylor polynomials of $\dfrac{1}{x^2+7}$

Laboratory Exercise 10.3

Interval of Convergence (CCH Text 10.2)

Name _____   Due Date _____

1. Use graphs as we did in Solved Problem 10.2 to estimate the interval of convergence of the Taylor series of $\ln x$ about $x = 2$, $x = 3$, and $x = 10$.

2. Based on your work in Part 1, conjecture the interval of convergence of the Taylor series of $\ln x$ about $x = n$ for $n > 0$.

3. Use graphs as we did in Solved Problem 10.2 to estimate the interval of convergence of the Taylor series of $\dfrac{1}{1+e^x}$ about $x=0$.

4. Use graphs as we did in Solved Problem 10.2 to estimate the interval of convergence of the Taylor series of the function $f(x) = \begin{cases} e^{-\frac{1}{x^2}} & \text{if } x \neq 0 \\ 0 & \text{if } x = 0 \end{cases}$ about $x=0$.

Remark: This example requires some careful thought to understand what is going on. Expand e^t about $t=0$ and then substitute $t=-1/x^2$ into the expansion. See CCH Text, Section 10.3.

Laboratory Exercise 10.4

Approximating π (CCH Text 10.3)

Name _____   Due Date _____

Example 4 of Section 10.3 of the CCH Text explains how to use the Taylor series for the arctangent function to obtain an approximation for π. It is noted there that while this method is very elegant, it is not practical for obtaining many digits of π. A better way, presented in Exercise 19, is to use Machin's formula:

$$\frac{\pi}{4} = 4\arctan\left(\frac{1}{5}\right) - \arctan\left(\frac{1}{239}\right).$$

Thus if $P(x)$ is the Taylor polynomial of degree n for $\arctan x$, then

$$\pi \approx 16P\left(\frac{1}{5}\right) - 4P\left(\frac{1}{239}\right).$$

1. Obtain the Taylor polynomial of degree 29 for $\arctan x$.

2. Set `Digits := 50`. Now, use your answer in Part 1 to approximate π.

3. How many digits of accuracy did you get from your answer in Part 2? (<u>Hint</u>: Compute `evalf(Pi - A);` where A is the approximation to π you obtained in Part 2.)

4. Let n be an odd integer. It can be shown that if $P(x)$ is the Taylor polynomial of degree n for $\arctan x$, then the error in approximating π using this method is no more than
$$\frac{4}{n+2}\left(\frac{4}{5^{n+2}} - \frac{1}{239^{n+2}}\right).$$
What degree Taylor polynomial is needed to obtain 100 digits of π using this method?

Solved Problem 10.3: The error in Taylor approximations (CCH Text 10.5)

Use the Taylor polynomial of degree 5 about $x = 0$ to approximate $\tan \frac{1}{2}$, and analyze the error.

Solution: We begin by defining the Taylor polynomial for $\tan x$ about $x = 0$ calling it p:

```
>p := convert(taylor(tan(x), x = 0, 7), polynom);
```

$$p := x + \frac{1}{3}x^3 + \frac{2}{15}x^5$$

Now approximate the value of $\tan \frac{1}{2}$ with

```
>evalf(subs(x = 1/2, p));
```
$$.5458333333$$

The maximum possible error in this approximation is determined by using the formula for the error-bound given in Section 10.5 of the CCH Text. In this case the error is no more than $\frac{M}{6!}\left(\frac{1}{2}\right)^6$, where M is the maximum value of the sixth derivative of $\tan x$ on the interval $[0, \frac{1}{2}]$. We compute and plot this derivative with the following commands. See Figure 10.3 for the graph.

```
>df6 := diff(tan(x), x$6);
```

$$df6 := 416(1 + \tan(x)^2)^2 \tan(x)^3 + 272(1 + \tan(x)^2)^3 \tan(x) + 32\tan(x)^5(1 + \tan(x)^2)$$

```
>df6 := expand(df6);
```

$$df6 := 1232\tan(x)^3 + 1680\tan(x)^5 + 720\tan(x)^7 + 272\tan(x)$$

```
>plot(df6, x = 0..1);
```

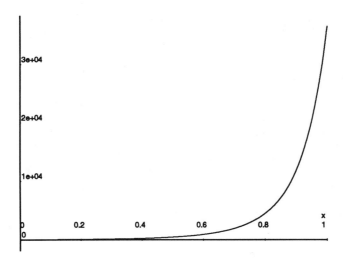

Figure 10.3: The sixth derivative of $\tan x$

Since the graph is increasing, it attains its maximum value on the interval $[0, \frac{1}{2}]$ at the right-hand endpoint, $\frac{1}{2}$. The following command will compute this value:

```
>evalf(subs(x = 1/2, df6));
```
$$441.6656546$$

Thus $M \approx 441.6656546$. Therefore the error is no more than

$$\frac{M}{6!}\left(\frac{1}{2}\right)^6 \approx \frac{441.6656546}{6!}\left(\frac{1}{2}\right)^6 \approx .009584758130$$

If we were to compute $\tan(1/2) - .5458333333$, we would see that the actual error is 0.0004691565. Notice that this is smaller than the predicted error, 0.009584758130. This illustrates the fact that $\frac{M}{(n+1)!}x^{n+1}$ gives an upper bound for the error, not the exact value of the error.

Laboratory Exercise 10.5

Approximating $\sec \frac{1}{2}$ (CCH Text 10.5)

Name _____   Due Date _____

1. Use the fifth-order Taylor polynomial for $\sec x$ about $x = 0$ to approximate $\sec \frac{1}{2}$.

2. Find the maximum error in your approximation above. Explain how you made your choice of M in the error formula.

3. Compute a decimal approximation of $\sec \frac{1}{2}$ using **evalf**. Compare this with the value in Part 1.

4. What is the smallest degree Taylor polynomial about $x = 0$ for $\sec x$ that can be used to approximate $\sec \frac{1}{2}$ to four digits accuracy?

Solved Problem 10.4: Fourier series (CCH Text 10.6)

Find the Fourier approximations of orders 1 through 4 for a periodic function that agrees with x on the interval $[-\pi, \pi]$. Graph $y = x$ together with the Fourier approximations.

Solution: We will define the Fourier polynomial just as it is defined on page 636 of the CCH Text. The coefficients of the polynomial are defined on page 637. You may wish to have your book open to those pages as you enter the following. We define the coefficients first:

```
>a0 := f -> (1/(2*Pi))*int(f(x), x = -Pi..Pi);
```

$$a0 := f \to \frac{1}{2} \frac{\int_{-\pi}^{\pi} f(x)\,dx}{\pi}$$

```
>a := k -> (1/Pi)*int(f(x)*cos(k*x), x = -Pi..Pi);
```

$$a := k \to \frac{\int_{-\pi}^{\pi} f(x)\cos(kx)\,dx}{\pi}$$

```
>b := k -> (1/Pi)*int(f(x)*sin(k*x), x = -Pi..Pi);
```

$$b := k \to \frac{\int_{-\pi}^{\pi} f(x)\sin(kx)\,dx}{\pi}$$

Notice that in the above definitions, $f(x)$ is an arbitrary function. This was done to make the definition as general as possible so that the Fourier polynomial, whose definition we now give, can apply to a wide variety of functions. We define the Fourier polynomial as a function of both f and the degree n,

```
>F := (f,n) -> a0(f) + sum(a(k)*cos(k*x), k = 1..n) +
       sum(b(k)*sin(k*x), k = 1..n);
```

$$F := (f, n) \to a0(f) + \left(\sum_{k=1}^{n} a(k)\cos(kx)\right) + \left(\sum_{k=1}^{n} b(k)\sin(kx)\right)$$

The function $F(f, n)$ produces the Fourier polynomial of degree n for a periodic function that agrees with $f(x)$ on the interval $[-\pi, \pi]$. In our case, $f(x) = x$.

To produce the graph, we first define f, gather together the functions we wish to plot in a set, then graph the set. See Figure 10.4. This may take a while to plot depending on the speed of your computer.

```
>f := x -> x;
```
$$f := x \to x$$

```
>Fset := {f(x), seq(F(f,n), n = 1..4)};
```
$$Fset := \{2\sin(x) - \sin(2x) + \tfrac{2}{3}\sin(3x), 2\sin(x) - \sin(2x), 2\sin(x),$$
$$2\sin(x) - \sin(2x) + \tfrac{2}{3}\sin(3x) - \tfrac{1}{2}\sin(4x), x\}$$

```
>plot(Fset, x = -2*Pi..2*Pi);
```

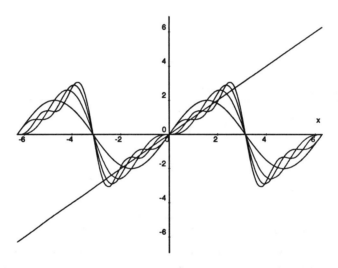

Figure 10.4: Fourier approximations for $y = x$

Remark: Rather than using the exact value of π as we did in defining the coefficients of the Fourier polynomial, sometimes it is helpful to use an approximation to π. To do this, we would first define

```
>p := evalf(Pi);
```

and then everywhere π appears in the definition of the coefficients, replace it with p. In Appendix II, we have provided code that does exactly this. You may use this file when doing Fourier approximations.

Laboratory Exercise 10.6

Fourier Approximations for $|x|$ (CCH Text 10.6)

Name _____   Due Date _____

1. Find the first five Fourier approximations to the periodic function that agrees with $|x|$ on the interval $[-\pi, \pi]$. (Hint: Use the file **FOURIER** given in Appendix II. See the remark at the end of Solved Problem 10.4.)

2. Attach a graph showing $|x|$ and its first five Fourier approximations. Explain the behavior of the graphs on and outside the interval $[-\pi, \pi]$.

Appendix I

Maple V Release 3 Reference and Tutorials

1. Introduction

The Maple V program comes with a fine set of manuals, but admittedly, they may not be accessible to first-time calculus students. This appendix is a concise reference to the major features of Maple V that often arise in the exercises in this book. Other features will be introduced as needed.

As with any software, Maple evolves, and there are some features of Release 3 that were not present in earlier ones, most of which will not be of concern to us.

Maple V is available for a very large number of different computer systems, each with its own nuances and peculiarities of operation. Each installation should have a "local expert." This may be your mathematics instructor, the person in charge of the math lab, or a fellow student. If you have a question about the way Maple works on your computer system, you should ask your local expert. The description we give here centers around the version of Maple V for Windows© although this pertains only to the interface, the way in which you, the user, interact with the program. If you are using Maple V on another computer such as a Macintosh, Amiga, or Unix-based computer, you will find that Maple acts in exactly the same way, although some features may not be available in some versions. For example, the ability to resize plot windows is not available under Maple V for DOS. The important thing is that, mathematically, Maple V Release 3 as described here will be the same on your computer.

Maple is a command-driven language. This means that commands must be typed and entered into the program. It is very important to note that Maple, like mathematics, is *case sensitive*. That is, Maple distinguishes between upper and lower case letters. Most Maple commands are entered in lower case, but occasionally upper case letters are used. You will also find that Maple commands, for the most part, are quite natural and reflect that aspect of mathematics that you want to perform.

When you first start Maple, you see Maple's logo, usually a picture of Sir Isaac Newton together with some copyright material. Once the program is loaded into your computer's memory, you will be presented with a blank screen together with a *prompt* with the cursor directly to its right. On the Windows© version, the prompt is the "greater than" symbol shown here

>

This is your signal that you may begin using Maple by entering commands and viewing and analyzing the output. Please note that the prompt symbol is not part of a Maple command and should *not* be typed.

Starting with Release 2 of Maple V, the output from Maple has been given in *display Postscript*©. This means that the output from Maple commands will look very much like the mathematics printed in your textbook. You will find this to be very convenient.[1]

Entering an expression in Maple involves customary syntax: addition (+), subtraction (−), division (/), exponents (^), and multiplication (*). (Multiplication *always* requires the use of *; that is, **2x** is not the same as **2*x**.) In Appendix III, we include a short list of common expressions as they appear in mathematics textbooks or when written and the corresponding correct Maple syntax for that expression.

If you need help during a session with Maple, just type a question mark, "?" followed by the name of the command with which you need help. For example, if you want help with the `factor` command, you would enter,

```
>?factor;
```

Maple will respond by displaying a Help screen. When you have finished reading the help file, close the screen in the manner customary on your computer system.

2. Some Conventions

When you type a command to Maple, Maple will respond, if it can, and then draw a horizontal line across the screen. This line is called a *separator*. It indicates that Maple has finished with the command you just entered and is waiting patiently for the next one. Separators can be turned on or off and your local Maple expert will show you how this can be done. We suggest that you leave separators turned on at least until you become accustomed to the way Maple behaves on your computer system. In the Maple commands we present in this book, we will use this convention of drawing the separator lines to indicate the end of a computation. For example, a typical command might be to evaluate π as a floating point (decimal) number. In this book we will indicate the command and the response from Maple as follows: (Remember: Do not type the prompt symbol.)

```
>evalf(Pi);
```
$$3.141592654$$

We will use this convention consistently, even though there are times when you may wish to turn separators off. If the response from a Maple command is a plot or the output

---

[1] This feature is not available in the DOS version.

is too long to list, we will draw the separator immediately after the command and either refer you to a figure containing the plot or describe what the output should be.

All Maple commands in this book will be denoted by using a typewriter style font `like this`. The output will be presented in a manner that is consistent with the way it will be displayed on your computer's monitor. After entering a command at a Maple prompt, press the **Return** key. If you are using a Macintosh, you must press the **Enter** key on the numeric keypad. We assume that you will press the **Return** key (or the **Enter** key) after each Maple command. No further mention of this will be made.

Most of the graphs given in this text were generated by Maple as Postscript© files, although a number of them are actual screen prints from the Windows© version of Maple. They reflect what you would see if you were to type the command into Maple.

3. Algebra Tutorial

Let's begin. Suppose we wish to factor the expression $x^2 + 3x - 4$. At a Maple prompt, we would type and enter (by pressing the **Return** key). Remember not to type the prompt symbol.

```
>factor(x^2 + 3*x - 4);
```

Notice the syntax used. The command `factor` is typed followed by an open parenthesis. Then the *argument* to the command, in this case, the polynomial $x^2 + 3x - 4$, is followed by a close parenthesis. Arguments to a command in Maple are almost always enclosed between open and closed parentheses. They represent the object that the command is supposed to work on to produce some output. Notice, in particular, the semicolon at the end of the command. The semicolon is called a *terminator*. It is your way of telling Maple that you are finished typing a command and that Maple should work on the problem and produce its output. Maple is not rude, and it will not interrupt you while you are entering a command. It will wait until you signal it with the terminator that you have finished. Terminators are very important in Maple's syntax. It is very common for new users to forget to use them and wonder why Maple is not responding. If you should happen to press the **Return** key without entering a terminator, just enter it on the next line, then press **Return** again. (The other terminator used in Maple is the colon (:). This is called the "silent terminator" because when it is used, Maple will work on your command but will not print its output on your screen. Sometimes, we are not interested in seeing the result of a calculation. We rarely will use the silent terminator in this book.)

The response from the above command is the factored form of $x^2 + 3x - 4$:
$$(x+4)(x-1)$$

To multiply out the factors above, we would enter

```
>expand((x-1)*(x+4));
```
$$x^2 + 3x - 4$$

Again, notice the use of the parentheses, the multiplication symbol, and the terminator. Now try factoring $x^2 + x + 1$ both by hand and by asking Maple.

Suppose we wish to solve the equation $x^2 + 3x - 4 = 0$ for x. From the above factors, we know that the solutions are $x = 1$ and $x = -4$. To have Maple solve this equation for us, we enter,

```
>solve(x^2 + 3*x - 4 = 0, x);
```
$$1, -4$$

The `solve` command in Maple can take up to two arguments. The first is the equation that is to be solved and the second, separated from the first by a comma, is the variable to be solved for. The above command asks Maple to "solve $x^2 + 3x - 4 = 0$ for x". If you do not specify an equation as the first argument to `solve`, Maple assumes you mean $= 0$. We will discuss solving equations in more detail later in this appendix.

To evaluate the quadratic $x^2 + 3x - 4$ at $x = 2$ by hand, we would substitute the value 2 for x in $x^2 + 3x - 4$ to get 6. Maple can do this for us in much the same way using its `subs`, short for *substitute*, command:

```
>subs(x = 2, x^2 + 3*x - 4);
```
$$6$$

If you read this command aloud, you will find that it is quite natural, "substitute $x = 2$ into $x^2 + 3x - 4$." This is true of most of Maple's commands.

The expression we have been working with may be given a name, or more correctly, be assigned to a variable, say p. This is done in Maple by entering,

```
>p := x^2 + 3*x - 4;
```
$$p := x^2 + 3x - 4$$

The symbol :=, the colon followed immediately by the equals sign with no intervening space, is called the *assignment operator*. Its purpose is to assign one expression to another which, in effect, makes them synonymous. Any reference to p is a reference to $x^2 + 3x - 4$. You can read the symbol := as "colon equals" or as "gets" as in "p gets $x^2 + 3x - 4$". The reader is encouraged now to try the following commands and observe their results.

```
>factor(p);
```

```
>solve(p = 0, x);
```

```
>subs(x = 2, p);
```

You should obtain exactly the same results you did above.

If you want to raise the quadratic $x^2 + 3x - 4$ to the power 6, you can enter

```
>(x^2 + 3*x - 4)^6;
```
$$(x^2 + 3x - 4)^6$$

Notice that Maple has returned to us the expression we entered. In order to see this sixth power expanded, we will use one of Maple's *internal variables*. These are variables that Maple uses for certain representations that it needs to make from time to time. Maple chooses them so that they are unlikely to be confused with the ones you use. The internal Maple variable we use here is denoted by the double quote symbol, ". Its value is the *most recently evaluated expression*. Thus, if you have just entered the above expression, " has the value $(x^2 + 3x - 4)^6$. Therefore if we enter

```
>expand(");
```

Maple will produce the expanded form of $(x^2 + 3x - 4)^6$. We will make frequent use of the double quote, ", in this book. It is quite useful and can save a lot of typing. Keep in mind that its value changes each time you execute a command. The value of " is the *most recently evaluated expression*.

If you have not changed the value of p, the same result could have been obtained with

```
>expand(p^6);
```

In either case Maple will produce the same long answer.

Another very useful command is **simplify**, which is used to simplify expressions. Sometimes, **simplify** and **expand** can have the same effect; for example, simplify(p^6) will produce the same result as in the previous example. These two commands, however, are really different. Enter

```
>p := (x^2 - 1)/(x*(x - 1));
```
$$p := \frac{x^2 - 1}{x(x - 1)}$$

and then execute the two commands, `simplify(p)` and `expand(p)`, to see the difference.

When using Maple as a calculator, keep in mind that Maple does exact rational arithmetic. This means that, unlike your scientific calculator, the result will be printed as a rational number reduced to lowest terms. For example, consider the following:

```
>(3+2*5)^2/7^3+1;
```
$$\frac{512}{343}$$

4. Functions, Expressions, and Constants

Maple distinguishes between functions and expressions and treats them differently. For example, $x^2 + 1$ is an expression while $f(x) = x^2 + 1$ defines a function.

We have seen examples of working with expressions in the last section. Note that the variable p we used there is also an expression and not a function. To work with functions in Maple, we first need to know how to indicate to Maple that a variable is to represent a function and not an expression. The method used is quite natural and reminisent of what we do in mathematics. Recall from your CCH Text, p. 2, that a function is a rule of correspondence which when applied to a value x, called the *argument* or *independent variable*, produces a unique value y. Using the notation $y = f(x)$ is one way of indicating this where f is the name we give to the rule itself. We sometimes say that f assigns the value x to the value y and write it as $f: x \to y$. Here y could be any expression we wish that defines how the assignment is to be made. For example, $f: x \to x^2 + 1$ is the function that assigns $x^2 + 1$ to a value x. We often see this written $f(x) = x^2 + 1$, but keep in mind that $f(x)$ represents the *value* of the function f at x, that is, the quantity that f has assigned to x. In this example, $f(2) = 5$ means that 5 is the value that f has assigned to 2. The notation $f: x \to y$ is nice because it separates the name of the function from its value. Maple uses a small variation of the notation to define its functions.

To define a function in Maple, we use the arrow notation, but instead of writing the name of the function to the left of a colon, we *assign* the function to the name it is to have. For example, enter the following at a Maple prompt:

```
>f := x -> x^2 + 1;
```

To create the arrow, type the minus sign immediately followed by the greater than symbol: `->`.

Maple will respond with
$$f := x \to x^2 + 1$$

Once this is done, you may treat f as a function in much the same way you would in mathematics. For example, we know that $f(2) = 5$. If you ask Maple for the value of $f(2)$, it will respond in like manner:

>f(2);
$$5$$

You may even evaluate f at variable expressions. For example, Maple easily determines $f(3 + t)$:

>f(3+t);
$$(3+t)^2 + 1$$

One of the important ideas in mathematics is that functions are entities unto themselves that may be manipulated in much the same way that we manipulate numbers. Maple makes it easy and convenient for us to do this and therefore strengthens our understanding of them.

Functions such as $f : x \to x^2 + 1$ are easily defined, but we know that functions can be quite complicated. For example, suppose that a function f is to have the value x^2 when $x < 0$ but is to have the value $-x$ when $x \geq 0$. How do we define such a function? In mathematics it is fairly simple:

$$f : x \to \begin{cases} x^2 & \text{if } x < 0 \\ -x & \text{if } x \geq 0 \end{cases}$$

In Maple, however, we cannot use the arrow notation because of the two cases involved. We can still define the function f but we must use an alternative approach: we define f as a *procedure*. To define the above function in Maple, we would enter

>f := proc(x) if x < 0 then x^2 else -x fi end;

In this definition, we are assigning to f a procedure for computing the value that f is to assign to x. The word proc is short for *procedure*. After proc we enclose the arguments of the function in parentheses, in this case x. Then we give a description of what f is to do in order to compute $f(x)$. In this example, we ask Maple to examine the value of x. If it is less than 0, we ask Maple to return x^2. On the other hand, the else part, if it is greater than or equal to 0, it is to return $-x$. The syntax requires that we end the if statement with `fi` (if spelled backwards). All procedures in Maple must end with an **end** statement.

Once this definition is made, you may treat f as you do any function defined in Maple. For example, for this function $f(-3) = 9$ while $f(3) = -3$. Maple will give these values if you now ask for them.

```
>f(-3); f(3);
```

(Notice how we put two statements on one line!) However, if you ask Maple to find $f(3+t)$, the result will be an error. This happens because Maple does not know whether $3 + t < 0$ or $3 + t \geq 0$. (There is a way to get around this with the **assume** command, but we shall not make use of it in this book. If you are interested, ask your local Maple expert.)

The procedure method of defining a function in Maple is presented here because you will see it several times in this book. You should also know that the arrow method of defining a function is actually an abbreviated version of the procedure method that became available starting with Release 1 of Maple V. In most instances, we will use the arrow method.

Maple knows about a large number of common functions (and some that are not so common). For example, Maple knows the standard trigonometric functions, their inverses, the exponential function and the logarithm functions. You will meet these as you work through this book.

Maple also knows the standard mathematical constants such as $\pi \approx 3.141592654$ and $e \approx 2.718281828$. It also knows the imaginary unit i and many others (one of which we will meet when we discuss the gamma function). While $\pm\infty$ are not constants, they have Maple representations nonetheless. These are represented in Maple as follows:

Mathematics	Maple
π	Pi (Capital P, small i)
e	E (Capital E)
i	I (Capital I)
$\pm\infty$	$\pm$infinity

5. Plotting Graphs

Being able to see a picture of the problem we are working on can be of immeasurable help. Maple's `plot` function is the essential function used to get two-dimensional graphs. The `plot` function can take several arguments, not all of which are required. Its underlying syntax is

$$\text{plot(f, h, v);}$$

or

$$\text{plot(f, h, v, ...);}$$

where `f` stands for the function, procedure, or expression or a set of functions, procedures, or expressions, or a list of points to be plotted, `h` represents the horizontal range, and `v` is the vertical range which is optional. The three dots in the second description represent various other options that are available when making a plot. We will use some of these options from time to time in this book.

A typical call to the plot function is

```
>plot(f(x), x = a..b);
```

where $f(x)$ is a real function in the variable x and `a..b` specifies the horizontal range over which $f(x)$ is to be plotted. If the range is not given, as in

```
>plot(f(x), x);
```

then Maple will plot $f(x)$ over its *default domain*, $-10 \leq x \leq 10$. In either case, the vertical range will be adjusted so the entire portion of the graph over the horizontal range will appear. As an example, the graph of $f(x) = x^3 - 5x + 10$ over the domain $[-3, 3]$ is the result of the following `plot` command and appears in Figure 0.1a. Notice that Maple displays units on both the horizontal and vertical axes. These are very helpful in understanding the graph.

```
>plot(x^3 - 5*x + 10, x = -3..3);
```

When Maple plots a function, it will open a new window containing the picture. If you wish, you may return to the worksheet, the screen on which you have been entering your Maple commands, and enter another `plot` command. Maple will then display your plot in a new window and you can then view the two plots together. This feature is available under Windows© and on Macintoshes and Amigas, but it is not available when using Maple under DOS. In the latter situation, you must close the plot window before plotting a new graph. See your local expert for more on this feature.

You may, however, plot more than one graph on the same set of axes. To do so, you create a *set* of functions, then plot the set. Sets in Maple are defined just as sets are defined in mathematics, by enclosing the elements in braces, { and }. For example, suppose we wish to plot x, x^2, and x^3 on the same axes over the domain $[-2, 2]$. The following command does this; see Figure 0.1b.

```
>plot({x, x^2, x^3}, x = -2..2);
```

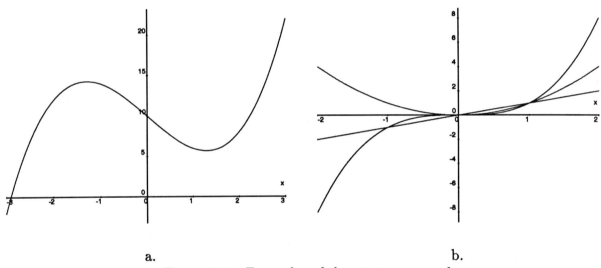

a. b.

Figure 0.1:   Examples of the `plot` command

Notice the set that appears as the first argument in the `plot` command. The set could have been defined separately as follows:

```
>s := {x, x^2, x^3};
```
$$s := \{x, x^2, x^3\}$$

Then the `plot` command below could have been entered. The graph will be the same as in Figure 0.1b.

```
>plot(s, x = -2..2);
```

Sets may also be manipulated in Maple just as they are in mathematics. The appropriate operations are union, intersect, and minus. You will see examples of these in the book.

Plotting individual points: In mathematics the coordinates of a point are represented as ordered pairs such as $(1, 2)$, $(-2, 1)$, or $(3, -1)$. In Maple, such points are represented as a two-entry list — the points are enclosed in square brackets instead of parentheses. Thus, in Maple these points would be entered as [1, 2], [-2, 1], [3, -1]. Points may be plotted in Maple in two ways. The first way is to include the points in a set, and then plot the set. However, we must use an option of the plot function. This option tells Maple to plot only points. See Figure 0.2 for the result of the following command.

>plot({[1, 2], [-2, 1], [3, -1]}, style = POINT);

In the second method, a single list is created with an even number of numeric entries, two numbers for each point to be plotted. Then the list is plotted. The following command produces a graph similar to the one in Figure 0.2.

>plot([1, 2, -2, 1, 3, -1], style = POINT);

It is important to note that POINT (not POINTS) is spelled in all capital letters. It is a syntax error to spell it otherwise. What occurs if the style = POINT option is not given? Try it and see. Also, try adding a horizontal range to the above plots.

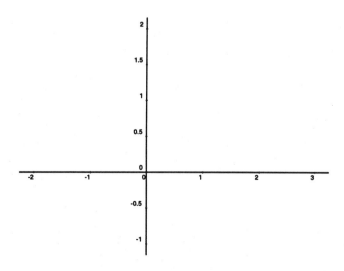

Figure 0.2:   A point plot

307

Plotting parametric curves: To plot the graph of a set of parametric equations, place the equations in a list together with the domain for the parameter. Then plot the list. For example, suppose we wish to plot the parametric equations, $x = \sin t$, $y = \sin(2t)$ on the interval $[-\pi, \pi]$. The following plot command shows how this is done; see Figure 0.3a.

```
>plot([sin(t), sin(2*t), t = -Pi..Pi]);
```

Try adding a horizontal range, then a horizontal and vertical range to the above plot command, and observe the results.

Plotting in polar coordinates: Plotting an equation in polar coordinates can be done in several ways in Maple, but we will discuss only one of them here. Let's consider the polar equation $r = f(\theta)$. Since polar equations are special cases of parametric equations with $x = \theta$ and $y = r = f(\theta)$, we can plot them using the method just described by adding a special option, coords = polar. As an example, consider the polar equation, $r = \sin(3\theta)$ (a three-leafed rose). The polar coordinate graph in Figure 0.3b is the result of the following command.

```
>plot([sin(3*t), t, t = 0..Pi], coords = polar);
```

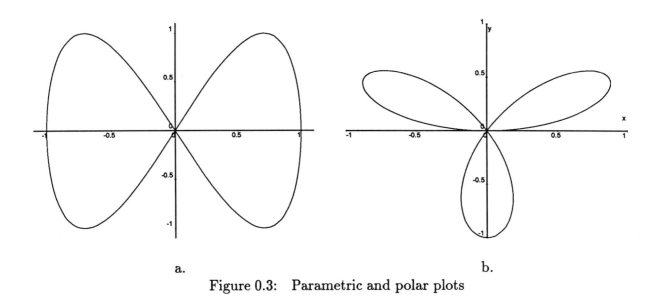

Figure 0.3: Parametric and polar plots

Note carefully the syntax used. The first entry in the list is the *radius*, r, and the second entry is the *angle*, θ. (We cannot type a θ so we have used t instead. You could have typed the word "theta" as in sin(3*theta) and Maple would have converted it to θ on your screen.)

Maple has a very powerful plots package, which is a collection of additional plotting routines. You gain access to this package by executing the following command:

>with(plots);

When this command is executed, Maple will list the routines in this package that are now available for use. This package offers another way to plot in polar coordinates using the polarplot routine. We shall not discuss it here, but you are encouraged to experiment with it if you like. Type ?polarplot at a prompt to learn how to use this function.

Plotting piecewise-defined functions: A piecewise-defined function is one that is defined in "pieces." We did an example in Section 4. Plotting such functions is done in much the same way as any other function, but some special care must be taken in order to avoid errors. Let's consider an example. Let g be defined as follows:

$$g(x) = \begin{cases} x - 1, & \text{if } -2 \leq x < 1 \\ \cos(x - 1), & \text{if } 1 \leq x \leq 6 \end{cases}$$

In Maple, we would enter

>g := proc(x) if x < 1 then x - 1 else cos(x - 1) fi end;
$g := \text{proc}(x) \text{ if } x < 1 \text{ then } x - 1 \text{ else } \cos(x - 1) \text{ fi end}$

Plot g over the domain $[-2, 6]$ and range $[-2, 2]$ using the following command. See Figure 0.4a.

>plot(g, -2..6, -2..2);

The syntax is important here. We **do not** want to plot $g(x)$. Doing so will result in an error. Give only the name of the procedure as the first argument to plot when functions are defined this way. Also, note that we do not supply names for the ranges. That is, we do not write x = -2..6, etc. In these cases, it is an error to do so.

Notice that in Figure 0.4a, the two "pieces" that make up the graph of g are connected. When Maple draws a graph, it does so by calculating a number of points that lie on the graph, and then connects them with small, curved segments called *splines*. Maple does not know that the graph of g is disconnected at $x = 1$, so it just connects the points as though it were connected. We can get a more accurate picture of the graph of g by using the style = POINT option and by increasing the number of points that Maple computes

309

before it produces the plot. This is done with another option, `numpoints`. The result of the following command appears in Figure 0.4b.

>`plot(g, -2..6, -2..2, numpoints = 200, style = POINT);`

Plotting equations that aren't functions: Suppose we wanted to obtain a graph of the equation $x^2 + y^2 = 1$, which we know represents the unit circle. The `plot` command will not do this for us because the equation does not define a function. If we solve the equation for y, then plot $y = \sqrt{1-x^2}$, we obtain only the top half of the circle. Since the equation

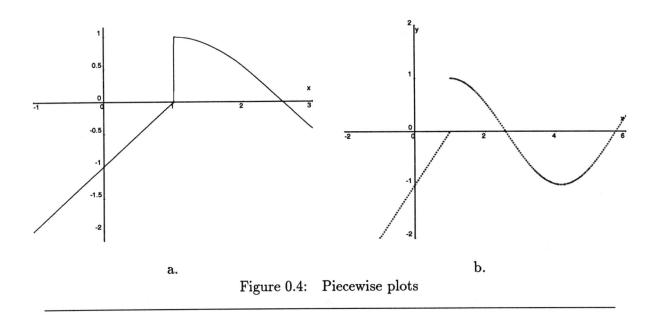

Figure 0.4: Piecewise plots

defines y *implicitly* as a function of x, we can resolve this problem using the `implicitplot` routine in the `plots` package. To use this routine, make sure that the `plots` package has been loaded into Maple's memory (see previous page). Now execute the following command whose graph appears in Figure 0.5.

>`implicitplot(x^2 + y^2 = 1, x = -1..1, y = -1..1);`

We shall make use of `implicitplot` again in this book.

6. A Plotting Tutorial

In this section we will show how to use the basic graphing skills learned in the last section. We will do this by using graphs to estimate the real solutions of $x^5 + 5x^2 - 1 = 0$. Let's begin by plotting the curve $y = x^5 + 5x^2 - 1$ with domain $[-3, 3]$ and range $[-10, 10]$; see Figure 0.6a.

```
>plot(x^5 + 5*x^2 - 1, x = -3..3, y = -10..10);
```

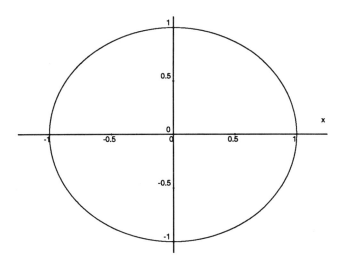

Figure 0.5: An implicitplot

The real solutions of the equation occur where the graph of $y = x^5 + 5x^2 - 1$ crosses the x-axis, and from Figure 0.6a it is clear that there are three solutions, one in the interval $(-2, -1)$, another in the interval $(-1, 0)$, and the third in $(0, 1)$. Let's estimate the positive zero, the one in the interval $(0, 1)$. If you are using a version of Maple that can be controlled with a mouse, move the mouse pointer to the point where the graph appears to cross the x-axis between 0 and 1 and click the left mouse button. (If you are using a Macintosh, just click the mouse button.) The coordinates of that point will appear on your plot window. (On the Windows© version, they appear in the lower left corner of the window.) Thus the positive root appears to be somewhat smaller than 0.5. Using this information, we "zoom in" on the plot near this point. That is, plot $y = x^5 + 5x^2 - 1$ again, but this time over the domain, $[0.4, 0.5]$. See Figure 0.6b.

```
>plot(x^5 + 5*x^2 - 1, x = 0.4..0.5);
```

This picture gives us a much clearer indication of what the positive solution is because it is clear that the graph crosses the x-axis between 0.44 and 0.45. Using this new information, we zoom in further on the plot. Execute the following command:

```
>plot(x^5 + 5*x^2 - 1, x = 0.44..0.45);
```

When you do, you will observe that the graph crosses the x-axis between 0.442 and 0.444. Thus the estimate that $x = 0.44$ is the positive solution of $x^5 + 5x^2 - 1 = 0$ would be correct to two decimal places.

To get the other real solutions graphically, we can proceed in a similar fashion by zooming in around those roots. You are encouraged to do so here. This method of zooming in on graphs to gain information about a function will be used at various places thoughout this book.

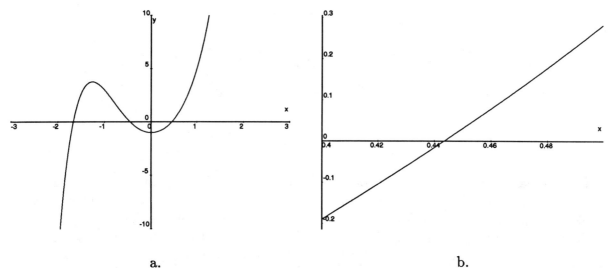

a.   b.

Figure 0.6: Zooming in on $y = x^5 + 5x^2 - 1$

7. Solving Equations Exactly

Suppose we wish to solve the equation

$$2x^4 + 9x^3 - 27x^2 = 22x - 48.$$

We can use Maple's `solve` command to try to find the exact solutions to equations. For this present example, enter

>solve(2*x^4 + 9*x^3 - 27*x^2 = 22*x - 48, x);
$$\tfrac{-3}{2}, 2, -\tfrac{5}{2} + \tfrac{1}{2}\sqrt{57}, -\tfrac{5}{2} - \tfrac{1}{2}\sqrt{57}$$

In this case Maple was able to find all the solutions of the equation, but this is not always the case. For example, in the last section, we learned that $x^5 + 5x^2 - 1 = 0$ has three real roots. We can ask Maple to solve this equation:

>solve(x^5 + 5*x^2 - 1 = 0, x);
$$RootOf(\_Z^5 + 5\_Z^2 - 1)$$

Maple is begging the question with its response. It is saying that whatever the roots are, they are the solutions of $x^5 + 5x^2 - 1 = 0$. (_Z is an example of the type of variable that Maple uses when it wants to express a concept. Maple uses the "underscore" before a capital letter because it is unlikely that a user will use the same construct.) But we already know this. In this case it will be necessary to approximate the solutions. Before we do so, however, it is necessary to discuss numerical precision in Maple.

8. Numerical Precision

Maple does exact rational arithmetic. This means that if the result of a computation is a rational number, Maple will express it as a ratio of integers reduced to lowest terms. Thus, if you enter the following command, Maple's response is a rational number in lowest terms.

>(13^2 - 9)/(2^4 + 3*7);
$$\frac{160}{37}$$

If you had entered this last computation on a calculator, the result would have been 4.324324324, the number of digits displayed depending on your calculator. Maple also does "floating point" calculations in many instances. For example, π is represented in Maple as `Pi`, but we can ask Maple to give us its numerical approximation by entering

```
>evalf(Pi);
```

`evalf` means "evaluate to a floating point number." When you execute this command, Maple will compute π to 10 significant digits. This is its internal default value: all such computations will contain 10 significant digits. This value can be overridden in two important ways. The first way is to do it *globally* by resetting the internal default. This is done by giving Maple's internal variable `Digits` (note the spelling; `Digits` must begin with a capital D) a new value. When Maple starts up, it is set to 10. Execute the following command.

```
>Digits := 20;
```
$$Digits := 20$$

Execute `evalf(Pi)` again, and you will see that Maple now has approximated π to 20 significant digits. Now reset `Digits` back to 10.

The second way to override the default number of significant digits is to do so *locally*. This is done directly in the `evalf` command as an option. For example,

```
>evalf(Pi, 25);
```

will approximate π correct to 25 significant digits. Caution must be used in extending the precision that Maple uses. An increase in the number of significant digits always increases its computation time. On some computers, this may cause an inordinate delay.

9. Solving Equations Approximately

If Maple fails to find exact solutions of an equation or if we wish only to find approximate solutions, we use Maple's `fsolve` command. The syntax for `fsolve` is almost like that for `solve` except that it has a few options that we will explore. Consider the fifth-degree equation $x^5 + 5x^2 - 1 = 0$ in the last section that Maple could not solve exactly. Let's try `fsolve`.

```
>fsolve(x^5 + 5*x^2 - 1 = 0, x);
```
$$-1.667977977, -.4513841740, .4433661724$$

Notice that Maple approximated all three real roots of the equation. (Recall that we obtained an approximation graphically for the positive real root in Section 6.) Maple will always give all real solutions when using the `fsolve` on polynomial equations. This is not true when nonpolynomial equations are involved, and we will give an example in a moment. Observe that the fifth-degree polynomial equation above has three real roots, so the other two roots must be complex and occur in conjugate pairs. `fsolve` can find these as well by specifying the option `complex` as the last argument:

```
>fsolve(x^5 + 5*x^2 - 1 = 0, x, complex);
```
$$-1.667977977, -.4513841740, .4433661724, .8379979894 + 1.514422832\ \text{I},$$
$$.8379979894 - 1.514422832\ \text{I}$$

As claimed, there are five roots, the last two being complex conjugates. Recall that the capital letter I is used by Maple to denote the imaginary unit i. See the table in Section 4.

Consider the equation $\sin x = 0$. It is impossible for Maple to find all the real solutions since are there infinitely many of them, all of which are integral multiples of π. If we enter

```
>fsolve(sin(x) = 0, x);
```

Maple will return only the answer 3.141592654. To find other solutions, we must specify an interval in which Maple is to look for the solution we seek. For example, if we want the second positive solution, which we know is 2π, we tell Maple to look in the interval $[6, 7]$ with

```
>fsolve(sin(x) = 0, x, 6..7);
```
$$6.283185307$$

For other functions whose zeros are sought, the interval in which to look may be determined by graphical methods as in Section 6. But if you ask a computer to plot a graph, how can you be sure that the screen displays all the points where the graph crosses the x-axis? In general, there is no simple answer to this question, but in the case of polynomials the following theorem can help.

The Confinement Theorem. Let $P(x) = a_n x^n + a_{n-1} x^{n-1} + \cdots + a_1 x + a_0$ be a polynomial with $a_n \neq 0$. Let M denote the largest of $|a_0|, |a_1|, \ldots, |a_n|$, and let $K = \dfrac{nM}{|a_n|}$. Then $P(x)$ has no zeros outside the interval $[-K, K]$.

Proof: Let $|x| > K$. Then for each $j = 0, 1, \ldots, n-1$ we have $|x| > \dfrac{n|a_j|}{|a_n|}$. Since $K > 1$, $|x|^{n-j} > \dfrac{n|a_j|}{|a_n|}$ also. Multiplying both sides of this last inequality by $\dfrac{|a_n x^j|}{n}$ gives

$$\frac{|a_n x^n|}{n} > |a_j x^j| \text{ for each } j = 0, 1, \ldots, n-1.$$

By the triangle inequality, $|\sum_{j=0}^{n} a_j x^j| \geq |a_n x^n| - \sum_{j=0}^{n-1} |a_j x^j|$. Combining the last two inequalities gives

$$|\sum_{j=0}^{n} a_j x^j| \geq |a_n x^n| - \sum_{j=0}^{n-1} |a_j x^j| > |a_n x^n| - \sum_{j=0}^{n-1} \frac{|a_n x^n|}{n} = 0.$$

This completes the proof.

If we apply this theorem to the polynomial $P(x) = x^5 + 5x^2 - 1$ we studied earlier, we see that $M = 5$ and $K = 25$. Therefore the graph of $P(x)$ cannot cross the x-axis outside the interval $[-25, 25]$. Plot the graph of $x^5 + 5x^2 - 1$ again, but this time specify the horizontal range as x = -28..28 and the vertical range as y = -20..20 and observe the result.

A Strategy for Solving Equations

Step 1: Use Maple's `solve` command to attempt to solve the equation exactly. If all the solutions are found, you do not need to proceed further.

Step 2: If exact solutions are not found, write the equation as $f(x) = 0$, plot $f(x)$, and isolate the zeros in small intervals.

Step 3: Use Maple's `fsolve` command on $f(x)$ on each of the intervals found in Step 2.

Step 4: Be alert to the possibility that zeros may be hiding outside the range of the plot. Plot $f(x)$ over different horizontal ranges to convince yourself that there are no more zeros or, in the case of a polynomial, use the Confinement Theorem.

10. Calculus

Some of the calculus operations that can be done in Maple are differentiating, integrating, finding limits, sums, and Taylor series. Other operations can be found in Maple's **student calculus** package and will be met in the Solved Problems. We give here some examples on using the basic routines to demonstrate their syntax.

Example: Find the derivative of $x^2 + 3x - 4$. Differentiation is done in Maple using its **diff** command. The variable of differentiation must be specified. Maple will treat all other variables as constants.

>diff(x^2 +3*x - 4, x);
$$2x + 3$$

Example: Find the derivative of $x^3 + 2xy + y^4$ with respect to x and with respect to y.

>diff(x^3 + 2*x*y + y^4, x);
$$3x^2 + 2y$$

>diff(x^3 + 2*x*y + y^4, y);
$$2x + 4y^3$$

Example: Find the second derivative of $2x^3 - 4x^2 + x - 1$. There are two ways to do this using **diff**. Either of the following two commands will give the correct answer.

>diff(2*x^3 - 4*x^2 + x - 1, x, x);
$$12x - 8$$

The second method makes use of the *repeat symbol* $.

>diff(2*x^3 - 4*x^2 + x - 1, x$2);

Note that in this case, the **x$2** means the same thing as ",x ,x". Clearly the repeat symbol is more useful when computing higher order derivatives.

Example: Find the indefinite integral of $x^2 + 3x - 4$ with respect to x; in other words, find the antiderivative of $x^2 + 3x - 4$. The appropriate Maple command here is **int**. As with **diff**, we must specify the variable of integration.

317

```
>int(x^2 + 3*x - 4, x);
```
$$\frac{1}{3}x^3 + \frac{3}{2}x^2 - 4x$$

If we want the definite integral of $x^2 + 3x - 4$ over an interval $[a, b]$, we will use the `int` command but this time the second argument will be the interval over which we wish to integrate specified as `x = a..b`. For example, if we want the definite integral of $x^2 + 3x - 4$ over $[0, 1]$, we will enter

```
>int(x^2 + 3*x - 4, x = 0..1);
```
$$\frac{-13}{6}$$

Example: Find the sum of the series $\sum_{n=1}^{\infty} 3^{-n}$. Maple's `sum` command easily handles problems such as these.

```
>sum(3^(-n), n = 1..infinity);
```
$$\frac{1}{2}$$

Example: Find the limit as x goes to infinity of $(1 + 1/x)^x$. For this example, we appeal to Maple's `limit` command. The first argument is the expression we wish to find the limit of, and the second argument is the limiting value:

```
>limit((1+1/x)^x, x = infinity);
```
$$e$$

11. Previous Expressions

In Section 3 we introduced one of Maple's internal variables, the double quote, ". See Section 3. Recall that the value of " is the most recently evaluated expression. It is possible to use this construct up to three times. Thus "" has the value of the expression before ", and """ the expression before that. The following sequence of commands illustrates their use.

```
>2 + 3;
```
$$5$$

" now has the value 5.

>" + x;
$$5 + x$$

"" now has the value 5 while " has the value $5 + x$.

>""*";
$$25 + 5x$$

""" has the value 5, "" is equivalent to $5 + x$, and " has the value $25 + 5x$. Can you determine the result of the following command?

>" - """^2 + "";

12. Trigonometry

Maple *always* uses radian measure for the trigonometric functions. Thus sin(30) in Maple means the sine of 30 radians, not 30 degrees. However, it is helpful to know how to convert from degrees to radians and vice versa. Conversions such as these are done with Maple's **convert** command. This command is very versatile and capable of many different kinds of conversions. We demonstrate its use here by converting degrees to radians and vice versa. We will use it in various other contexts in this book.

If you enter

>convert(x, degrees);

Maple will respond with
$$180\frac{x \; degrees}{\pi}$$

in keeping with Maple's inherent nature to return symbolic results. To see the result as a number, use **evalf** in conjunction with **convert**. Thus to convert $\pi/7$ radians to degrees, enter

>evalf(convert(Pi/7, degrees));
$$25.71428571 \; degrees$$

To convert x degrees to radians, use the general form,

>`convert(x*degrees, radians);`

Notice, in particular, the multiplication symbol between "x" and "degrees." Thus to convert 90 degrees to radians, enter

>`convert(90*degrees, radians);`
$$\frac{1}{2}\pi$$

---

13. Naming Variables, Expressions, and Functions

In mathematics it is common to use letters such as f and g to represent functions and x and y to represent variables, although we are by no means limited to these. Longer strings are possible and often used. Examples are log and sin for the logarithm and sine functions, respectively. Longer strings are also possible in Maple. So, for example, you can name a function (or variable or expression) you are working with "joe" or "mary" if you wish. There are some restrictions that must be observed, however. A Maple variable starts with a letter and can be followed with up to 498 letters, digits, or underscores. Typical variable names aren't close to the maximum, but they should be convenient to type and easy to remember. The following are some possible valid names that can be used in Maple:

R2D2, Formula_10, df, ddf, expr1, TaylorPoly.

Here are some cautions to observe when naming variables:

a. Don't use names that Maple already knows such as `log`, `sin`, or `exp`, or the name of any Maple command.

b. Don't use spaces, periods, or commas in the name you wish to use. (Actually, you can use these, but we are not going to tell you how.) The only character other than a letter or digit that you should at all consider is the underscore character. It is possible to begin a name with the underscore character as in _X but this practice should be avoided. Maple uses such "underscore variables" in its internal programming, and they sometimes appear in system-generated symbols or constants. See Caution d below.

c. *Maple is case sensitive.* For example, `abc` is considered by Maple to be different from `AbC`. Thus if you give a variable the name `abc`, Maple will not recognize it as `AbC`.

d. When considering a longer name, try to use capital letters whenever possible. This will avoid a possible conflict with a Maple internal name. For example, if you want to name a polynomial "taylor" to remind you of its origin, don't do it. `taylor` is the name of a Maple command that produces a Taylor series. Use TAYLOR or Taylor or some other such name instead.

The later releases of Maple V have tightened up considerably on the problem mentioned in Caution a. In Release 3, Maple will not let you assign a value to an internal Maple constant or other internal variable.

14. Other Data Structures

The three Maple data structures that are important for this book are expression sequences, lists, and sets.

An *expression sequence* is just an ordering of Maple objects separated by commas. Thus, 1, 1, 2, 3 is an expression sequence. Expression sequences sometimes result from a Maple command. For example, if we ask Maple to solve the cubic equation, $x^3 + 8x^2 + x - 42 = 0$, the result will be 2, −3, −7, an expression sequence. Expression sequences also are the result of the seq command which is explained in those parts of this book where it is used. They can be assigned names, and there is an empty sequence referred to as NULL. It is also important to remember that expression sequences preserve the order of the data.

A Maple *list* may be created by enclosing an expression sequence in square brackets, [and]. For example,

[1, 2, 3], [x^2, sin(x), -7, f(t)], and [[-2, 1], [x, x^2, -3*x]]

are all lists. Like expression sequences, lists preserve the order of the data in them.

A Maple *set* is written as a sequence enclosed by braces (curly brackets), { and }, which is the usual mathematical notation for finite sets. Sets in Maple do not preserve the order of the data in them, nor can multiple items be defined in a set. Sets may be manipulated in Maple with the set operations union, intersect, and minus. Thus if A and B are sets, then A union B means $A \cup B$, A intersect B means $A \cap B$, and A minus B means $A - B$. Sets occur in several places in this book, most notably when plotting graphs.

15. Unassigning Variables

Sometimes it will be convenient to have Maple "forget" the value assigned to a variable. This can be done in two ways. If just a single variable is involved, assign the variable its own name. This is done as follows:

>p := 'p';

$$p := p$$

Notice the use of the single quotes around **p**. The other method is global, that is, by returning Maple to the state it is in when first started up. This is done with the `restart` command:

>`restart;`

Executing this command will erase all of Maple's memory, including all variables and packages. Use this command with caution.

16. Line Editing

Generally speaking, you may edit your commands by using the four arrow keys on the right-hand side of your keyboard. These allow you to move up, down, right, and left. The delete and backspace keys can then be used to alter the line's contents. For more specific information on this topic for your computer system, see your local expert.

17. Some Common Errors

1. Forgetting the terminator.

2. Forgetting the left and/or right parentheses, forgetting to use the multiplication symbol.

3. Misspelled commands. E.g., `slove` instead of `solve`.

4. Unknowingly using a variable that previously has been assigned.

Appendix II

Optional Files for Use in Maple V Release 3

The procedures given in this appendix are designed to produce either graphical displays or to yield a specific calculation quickly. You only need to understand what a procedure is supposed to do in order to use it. You do not need to understand the code. We think the displays produced or the calculations that result are worth the time it takes to enter and save these routines.

These procedures are designed for use with Release 3. The minor changes needed to make them work for earlier releases will be mentioned.

While it is possible to enter these procedures directly in a Maple session, it is better to use a text editor. An editor makes it is easy to enter and save these files and to correct any mistakes you may make in entering the code. The Maple V program comes with an easy-to-use editor, called MapleEdit, for just such a purpose. Ask your local expert how to use MapleEdit on your computer system. He or she also can explain to you how to save these files to a floppy disk using file-naming conventions on your computer system. You can then bring them with you to the lab when you are working on an assignment.

Be careful to type the expressions <u>exactly</u> as they appear. Take special care to distinguish square brackets, braces (curly brackets), and parentheses. Don't omit commas, semicolons, or colons, and be sure to distinguish ":=" from "=". Each procedure terminates with a colon. You may, if you wish, end them with a semicolon. If you do, then when the procedure is entered into Maple, it will be printed out on your monitor.

1. Difference Quotient

We will begin with a very easy procedure we call DQ for <u>D</u>ifference <u>Q</u>uotient. Once a function has been defined in Maple, DQ will give its difference quotient. Enter the following code and save the file with the name DQ.

```
DQ := (f, x, h) -> ((f(x + h) - f(x))/h:
```

To use this routine you first need to read it into Maple's memory. You do this with Maple's **read** command. For example,

```
>read'DQ';
```

Several items are worth noting here. First, the file name *must* be surrounded with back single quotes. The back single quote is usually found on the left side of the keyboard just above the tab key. If you do not do this, Maple will return an error. The second item to note is that you may have to prefix the file name with a path indicating where the file is located. For example, if you are using a computer that uses DOS or Windows© and your file is on a floppy disk in drive a, then you would enter

```
>read'a:DQ';
```

Now that DQ is in Maple's memory, you may use it just as you would any other Maple command. The first argument to DQ must be a function, not an expression. The second argument is the point at which the difference quotient is to be computed, and the third argument is an increment. For example, to find the difference quotient of a function f at $x = 2$ when $h = 0.1$, enter

```
>DQ(f, 2, 0.1);
```

To determine the derivative of a function at a point $x = a$ using the definition, enter

```
>limit(DQ(f, a, h), h = 0);
```

In the last example, you may need to simplify the result of DQ before applying the limit. Try applying DQ to the secant function; then compute the limit of the result before and after simplifying the result.

An easy way to obtain an approximate graph of the derivative of a function f is to plot DQ(f, x, h) with a small value of h. For example, we might enter

```
>plot(DQ(f, x, 0.1), x);
```

to produce an approximation to the graph of f'.

2. Numerical Routines

2.1. Newton's method

```
NEWTON := proc(f, x0, n)
    local i, x, z, iter:
    df := diff(f(z), z):
    x := evalf(x0):
    i := 1:
    while i <= n do
        iter := evalf(x − f(x)/subs(z = x, df)):
        i := i + 1:
        x := iter:
    od:
    iter;
end:
```

The above procedure is used to apply Newton's method for approximating the zeros of a function. The first argument must be a function, the second argument is the starting point for the iteration, and the third argument is the number of iterations. See Solved Problem 5.5 for more information on Newton's method. The following procedure will produce a graphical display of the Newton iteration. It seems reasonable to include this routine here rather than with the graphical procedures in the next section. Save both this and the previous routine using the file name **Newton**.

```
NEWTONPIC := proc(f, x0, n)
    local set, j, list, lines, curve, z:
    z := evalf(x0):
    set := z:
    list := z, 0, z, f(z):
    for j from 1 to n do
        set := set, NEWTON(f, x0, j):
        list := list, NEWTON(f, x0, j), 0, NEWTON(f, x0, j), f(NEWTON(f, x0, j)):
    od:
    with(plots)[display]:
    list := [list]:
    lines := plot(list, style = LINE):
    curve := plot(f(x), x = min(set)−1..max(set)+1):
    display({lines,curve}, x = min(set)−1..max(set)+1);
end:
```

2.2. Euler's method

Chapter 9 describes the use of Euler's method for numerically solving differential equations. Save this file using the name EULER.

```
EULER := proc(expr, varlist, initlist, h0, n)
    local i, xk, yk, h, varA, varB, f, set, pointlist, tlist:
    global pointset:
    flag := 0:
    if nargs > 5 then flag := 1: fi:
    xk:=evalf(initlist[1]): yk:=evalf(initlist[2]):
    varA := varlist[1]: varB := varlist[2]:
    f := unapply(expr, varA, varB):
    h := h0:
    tlist := initlist:
    if flag = 0 then print(xk, yk); fi:
    for i from 1 to n do
        yk := yk + h*f(xk,yk):
        xk := xk + h:
        tlist := tlist, [xk,yk]:
        if flag = 0 then
            print(xk, yk);
        fi:
    od:
    pointlist := [tlist]: set:= tlist:
    pointset := pointlist union set:
    tlist := NULL:
end:
```

Remark: If you are using a release earlier than Release 3, do not enter the global statement in line 3 above.

2.3. Fourier approximation

Solved Problem 10.4 deals with Fourier approximations. The following routine can be useful. See that problem for details. Enter the following lines exactly as they appear here and save them as a single file called FOURIER. The function you will use is on the last line.

```
p := evalf(Pi):
a0 := f -> (1/(2*p))*int(f(x), x = -p..p):
a := k -> (1/p)*int(f(x)*cos(k*x), x = -p..p):
b := k -> (1/p)*int(f(x)*sin(k*x), x = -p..p):
FOURIER := (f, n) -> a0(f) + sum(a(k)*cos(k*x), k = 1..n) + sum(b(k)*sin(k*x), k = 1..n):
```

2.4. Riemann sums

Riemann sums are used in this book in Chapters 3 and 7. The routines given here may be used when working through those chapters. In many instances, they provide for greater speed of computation, depending, of course, on your computer system. They were named to coincide with those in Chapter 7 of the CCH Text. You should save all of these procedures as a single file called **SUMS**. Ask your local expert on the method for doing this on your system.

```
LEFT := proc(expr, intrvl, n)
    local a, b, h, s, i, var;
    var := op(1, intrvl):
    a := op(1, op(2, intrvl)): b := op(2, op(2, intrvl)):
    h := (b − a)/n:
    s := 0:
    for i from 0 to n−1 do
        s := s + evalf(h*subs(var = a + i*h, expr)):
    od:
    s;
end:


RIGHT := proc(expr, intrvl, n)
    local a, b, h, s, i, var;
    var := op(1, intrvl):
    a := op(1, op(2, intrvl)): b := op(2, op(2, intrvl)):
    h := (b − a)/n:
    s := 0:
    for i from 1 to n do
        s := s + evalf(h*subs(var = a + i*h, expr)):
    od:
    s;
end:


MID := proc(expr, intrvl, n)
    local a, b, h, s, i, var;
    var := op(1, intrvl):
    a := op(1, op(2, intrvl)): b := op(2, op(2, intrvl)):
    h := (b − a)/n:
    s := 0:
    for i from 1 to n do
        s := s + evalf(h*subs(var = a + ((2*i − 1)/2)*h, expr)):
    od:
    s;
end:
```

```
TRAP := proc(expr, intrvl, n)
    local s;
    s := (LEFT(expr, intrvl, n) + RIGHT(expr, intrvl, n))/2:
    s;
end:

SIMP := proc(expr, intrvl, n)
    evalf((2*MID(expr, intrvl, n) + TRAP(expr, intrvl, n))/3);
end:
```

These procedures are easy to use, and they all have the same syntax which we illustrate using LEFT.

$$\texttt{LEFT(expr, var = a..b, n)};$$

The first argument, **expr**, is the expression whose integral is to be approximated; the second argument, **var = a..b**, is the interval over which we want to approximate **expr** where **var** is the variable in **expr**. Lastly, **n** is the number of subintervals to be used in this approximation. Example:

```
>LEFT(sin(t^2), t = 0..2.5, 200);
                        .4307901870
```

---

3. Graphical Routines

In Chapters 3 and 7 of this book, we will use some of Maple's routines for displaying graphical representations of the left-hand, right-hand, and midpoint Riemann sums. The following procedure complements those routines in that it produces a graphical representation of the trapezoidal rule. Its syntax is exactly the same as that for **leftbox** or **rightbox** in the **student calculus** package. Enter both of the following procedures carefully and save them as a single file called **TRAPBOX**.

```
trpbox := proc(f, h, a, b, i)
    local xleft, xright, yleft, yright;
    yleft := f(a + i*h):
    yright := f(a + (i+1)*h):
    RETURN(a + i*h, 0, a + i*h, yleft, a + (i + 1)*h, yright, a + (i + 1)*h, 0);
end:
```

```
TRAPBOX := proc(expr, intrvl, n)
    local a, b, x, var, g, boxes, FirstPlot, SecondPlot;
    global approx:
    var := op(1, intrvl)
    g := unapply(expr, var):
    a := op(1, op(2, intrvl)): b := op(2, op(2, intrvl)):
    h := (b - a)/n:
    boxes := NULL:
    for i from 0 to n−1 do boxes := boxes, trpbox(g, h, a, b, i): od:
    boxes := [boxes]:
    FirstPlot := plot(g(x), x = a..b):
    SecondPlot:= plot(boxes, x = a..b, style = LINE):
    approx := {plot(boxes, x = a..b)}:
    with(plots)[display]:
    display({FirstPlot, SecondPlot});
end:
```

Remark: If you are using a release earlier than Release 3, do not enter the global statement in line 3 above.

As an example, the following command will produce an illustration of the trapezoidal rule for $f(x) = x^2$ on $[0, 2]$.

```
>TRAPBOX(x^2, x = 0..2, 9);
```

A convenient feature of the routine is that it produces a plot structure, called approx, that will display the trapezoidal boxes without the graph of the function. For example, after executing the previous command, enter the following to see the graph.

```
>display(approx);
```

This next procedure provides a graphical display of Simpson's rule. Its syntax is the same as that for TRAPBOX *except that the value of n must be even.* If it is not, the procedure will return an error message. Thus if you enter

```
>SIMPBOX(sin(x), x = -Pi..Pi, 6);
```

---

you get the graph in Figure 0.7a. However, you may provide an optional fourth argument in the form x = c..d. This argument provides a larger (or smaller) view window for the graph. The picture in Figure 0.7b is the result of the following command.

```
>SIMPBOX(sin(x), x = -Pi..Pi, 6, x = -4.5..4.5);
```

Like TRAPBOX, SIMPBOX produces the plot structure **approx** that displays the Simpson's boxes without the graph. For example, after executing the previous command, enter

```
>display(approx);
```

SIMPBOX := proc(expr, intrvl, n)
    local graph, L, s, p, q, f, Rt, Lt, h, i, m, a, b, A, B, var, z, c, d:
    global approx:
    if n mod 2 <> 0 then ERROR(n, 'is not an even integer') fi:
    a := op(1, op(2, intrvl)): b:= op(2, op(2, intrvl)):
    if nargs > 3 then
        c := op(1, op(2, args[4])): d := op(2, op(2, args[4])):
        c := evalf(c): d := evalf(d):
    else
        c := evalf(a): d := evalf(b):
    fi:
    h := (b − a)/n:
    var := op(1, intrvl):
    f := unapply(expr, var):
    for i from 0 by 2 to n−2 do
        Lt := NULL: Rt := NULL: Md := NULL:
        A := a + i*h: B := a + (i + 2)*h:
        m := a + (i + 1)*h:
        Lt := A, 0, A, f(A): Lt := [Lt]:
        Md := m, 0, m, f(m): Md := [Md]:
        Rt := B, 0, B, f(B): Rt := [Rt]:
        q := f(A)*((var − m)*(var − B))/((A − m)*(A −B)):
        q := q + f(m)*((var − A)*(var − B))/((m − A)*(m − B)):
        q := q + f(B)*((var − A)*(var − m))/((B − m)*(B −A)):
        p.i := plot({Rt, Lt, Md, q}, var = A..B):
    od:
    graph := plot(f(z), z = c..d):
    with(plots)[display]:
    i := 'i': L := [seq(2*i, i = 0..n/2−1)]: i := 'i':
    s := {graph, seq(p.i, i = L)}: approx := {seq(p.i, i = L)}:
    display(s, x = c..d);
end:

Remark: If you are using a release earlier than Release 3, do not enter the global statement in line 3 above.

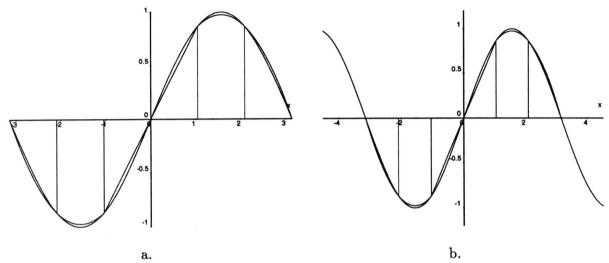

a.   b.

Figure 0.7:   Examples of the SIMPBOX command

Appendix III

Common Maple Syntax

Text or handwritten notation	Maple syntax
Addition $a + b$	a + b
Subtraction $a - b$	a - b
Multiplication ab	a*b
Division a/b or $\frac{a}{b}$	a/b
Mixed operations $a(b+c)$	a*(b + c), not a(b + c)
Mixed operations $\frac{a}{b+c}$	a/(b+c), not a/b + c
Mixed operations $\frac{a+b}{c+d}$	(a + b)/(c + d), not a + b/c + d
Mixed operations $a + \frac{b}{c+d}$	a + b/(c + d)
Exponentiation a^n	a^n
Negative exponents a^{-n}	a^(-n), not a^-n
Fractional exponent $a^{m/n}$	a^(m/n), not a^m/n
Square root $\sqrt{a}$ or $a^{1/2}$	sqrt(a) or a^(1/2)
Absolute value $|x|$	abs(x)
The number π	Pi (Capital P, small i)
The number e	E (Capital E)
Factorials $n!$	n!
$\sin x$, $\sin(x)$	sin(x), not sin x
$\sin^2 x$	sin(x)^2, not sin^2(x) nor sin^2x
$\sin x^2$, $\sin(x^2)$	sin(x^2), not sin x^2
Exponential function e^x	exp(x) or E^x, not e^x
$\ln x$, $\ln(x)$	ln(x) or log(x), not ln x nor log x
$\log_{10} x$, $\log_{10}(x)$	log[10](x)
Imaginary unit i	I (Capital I)

FUNCTIONS

$f(x) = x^3 - 5x + 10$
    f := x -> x^3 - 5*x + 10
    not f(x) = x^3 - 5*x + 10

$g(x,y) = 2x^2 + 4y - \sin(xy)$
    g := (x, y) -> 2*x^2 + 4*y - sin(x*y)

You may use any variable(s) you wish when defining a function. The variable(s) that appear on the left-hand side of the arrow should also appear on the right-hand side. In general:

$f(x) = \text{expr(ession) in } x$ `f := x -> expr in x`

$g(t) = \text{expr in } t$ `g := t -> expr in t`

$h(x,y) = \text{expr in } x \text{ and } y$ `h := (x, y) -> expr in x and y`

Index of Solved Problems

Chapter 1

Domains, ranges, and zeros of functions (CCH Text 1.1) 1
Testing exponential data (CCH Text 1.3) 9
Powers versus exponentials (CCH Text 1.4) 17
Inverses of functions (CCH Text 1.5) 23
Properties of logarithms from graphs (CCH Text 1.6) 33
Approximating the number e (CCH Text 1.7) 37
Periodicity of trigonometric functions (CCH Text 1.10) 53
Rational functions (CCH Text 1.11) 59
Vertical asymptotes (CCH Text 1.11) 60

Chapter 2

Calculating velocities (CCH Text 2.1) 65
Average rates of change (CCH Text 2.2) 73
Instantaneous rates of change (CCH Text 2.2)) 75
The derivative function (CCH Text 2.3) 87
Calculating limits graphically (CCH Text 2.7) 93

Chapter 3

Measuring distance (CCH Text 3.1) 97
Calculating Riemann sums (CCH Text 3.2) 103
Limits of Riemann sums (CCH Text 3.2) 106

Chapter 4

Difference quotients and derivatives (CCH Text 4.2) 123
The product rule (CCH Text 4.4) 127
Implicit differentiation (CCH Text 4.8) 131

Chapter 5

Maxima and minima (CCH Text 5.1) 137
Critical points and extrema (CCH Text 5.1) 140
Inflection points (CCH Text 5.2) 145
Welding boxes (CCH Text 5.6) 157
Newton's method (CCH Text 5.7) 165

Chapter 6

Families of antiderivatives (CCH Text 6.3) 169

Chapter 7

- Integrating with Maple (CCH Text 7.1) 175
- Implementing the right-hand rule (CCH Text 7.6) 179
- Implementing the midpoint rule (CCH Text 7.6) 180
- Calculating improper integrals (CCH Text 7.8) 191
- Approximating improper integrals (CCH Text 7.9) 199
- Approximating improper integrals II (CCH Text 7.9) 205

Chapter 8

- An oil slick (CCH Text 8.1) 211
- Area and center of mass (CCH Text 8.2) 213
- Arc length, area, and volume (CCH Text 8.2) 219
- From the Earth to the moon (CCH Text 8.3) 227
- Rainfall in Anchorage (CCH Text 8.6) 231

Chapter 9

- Families of solutions (CCH Text 9.1) 237
- Slope fields (CCH Text 9.2) 241
- Euler's method (CCH Text 9.3) 245
- A tank of water (CCH Text 9.6) 249
- Terminal velocity (CCH Text 9.6) 255
- Population growth with a threshold (CCH Text 9.7) 259
- Spider mites and lady bugs (CCH Text 9.8) 265
- Springs and second-order equations (CCH Text 9.10) 269

Chapter 10

- Approximating $\cos x$ (CCH Text 10.1) 275
- Intervals of convergence (CCH Text 10.2) 283
- The error in Taylor approximations (CCH Text 10.4) 289
- Fourier series (CCH Text 10.5) 293

Index of Laboratory Exercises

Chapter 1

Zeros, Domains, and Ranges (CCH Text 1.1)......................7
Fitting Exponential Data (CCH Text 1.3)........................13
U.S. Census Data (CCH Text 1.3)................................15
Powers versus Exponentials (CCH Text 1.4)......................21
The Inverse of a Function (CCH Text 1.5).......................27
Restricting the Domain (CCH Text 1.5)..........................29
The Inverse of an Exponential Function (CCH Text 1.5)..........31
Seeing Log Identities Graphically (CCH Text 1.6)...............35
Approximating e (CCH Text 1.8)...............................41
Seeing Log Identities Graphically II (CCH Text 1.7)............43
Growth Rates of Functions (CCH Text 1.7).......................45
A Graphical Look at Borrowing Money (CCH Text 1.8).............47
Shifting and Stretching (CCH Text 1.9).........................49
Equations Involving Trig Functions (CCH Text 1.10).............55
Inverse Trig Functions (CCH Text 1.10).........................57
Asymptotes (CCH Text 1.11).....................................63

Chapter 2

Average Velocity (CCH Text 2.1)................................69
Falling with a Parachute (CCH Text 2.1)........................71
Slopes and Average Rates of Change (CCH Text 2.1)..............79
Tangent Lines and Rates of Change (CCH Text 2.2)...............81
The Derivative of the Gamma Function (CCH Text 2.2)............89
The Meaning of the Sign of f' (CCH Text 2.3).................89
Recovering f from f' (CCH Text 2.5)........................91
Using Graphs to Estimate Limits (CCH Text 2.7).................95

Chapter 3

An Asteroid (CCH Text 3.1).....................................99
A Falling Water Table (CCH Text 3.1)..........................101
Calculating Riemann Sums (CCH Text 3.2).......................109
Limits of Riemann Sums (CCH Text 3.2).........................111
Estimating Integrals with Riemann Sums (CCH Text 3.2).........113
Riemann Sums and the Fundamental Theorem (CCH Text 3.4).......115
Calculating Areas (CCH Text 3.3)..............................119
The Average Value of a Function and the Fundamental Theorem
    (CCH Text 3.4)..121

Chapter 4

Difference Quotients and the Derivative (CCH Text 4.2) 125
The Quotient Rule and the Chain Rule (CCH Text 4.5) 129
Implicit Differentiation (CCH Text 4.8) 135

Chapter 5

Local Extrema (CCH Text 5.1) 143
Inflection Points (CCH Text 5.2) 147
The George Deer Reserve (CCH Text 5.2) 149
The Meaning of the Signs of f' and f'' (CCH Text 5.2) 151
Families of Curves (CCH Text 5.3) 155
Building Boxes (CCH Text 5.6) 159
Building Fuel Tanks (CCH Text 5.6) 161
Roads (CCH Text 5.6) 163
Newton's Method (CCH Text 5.7) 167

Chapter 6

Antiderivatives of $\arctan x$ (CCH Text 6.3) 171
An Antiderivative of $\sin(x^2)$ (CCH Text 6.3) 173

Chapter 7

Integrating with Maple (CCH Text 7.1) 177
Implementing the Left-Hand Rule (CCH Text 7.6) 183
Implementing the Trapezoidal Rule and Simpson's Rule
    (CCH Text 7.7) 187
The Trapezoidal Rule with Error Control (CCH Text 7.7) 189
Evaluating Improper Integrals (CCH Text 7.8) 195
The Gamma Function Revisited (CCH Text 7.8) 197
Approximating Improper Integrals (CCH Text 7.9) 203
Approximating Improper Integrals II (CCH Text 7.9) 209

Chapter 8

An Oil Spill in the Ocean (CCH Text 8.1) 215
The Center of Mass of a Sculpture (CCH Text 8.1) 217
Arc Length and Volume (CCH Text 8.2) 221
Arc Length and Limits (CCH Text 8.2) 223
Ratio of Arc Length to Area (CCH Text 8.2) 225
From the Earth to the Sun (CCH Text 8.3) 229
SAT Scores (CCH Text 8.6) 235

Chapter 9

    Families of Solutions (CCH Text 9.1) 239
    Slope Fields (CCH Text 9.2) 243
    Estimating Function Values with Euler's Method (CCH Text 9.3) 247
    A Leaky Baloon (CCH Text 9.5) 251
    Baking Potatoes (CCH Text 9.5) 253
    Drag and Terminal Velocity (CCH Text 9.6) 257
    Comparing Population Models (CCH Text 9.7) 263
    Foxes and Hares (CCH Text 9.8) 267
    A Damped Spring (CCH Text 9.11) 273

Chapter 10

    Taylor Polynomials and the Cosine Function: The Expansion Point
        (CCH Text 10.1) 279
    Taylor Polynomials and the Cosine Function: The Degree
        (CCH Text 10.2) 281
    Interval of Convergence (CCH Text 10.2) 285
    Approximating π (CCH Text 10.3) 287
    Approximating $\sec\frac{1}{2}$ (CCH Text 10.5) 291
    Fourier Approximations for $|x|$ (CCH Text 10.6) 295

NOTES

NOTES

NOTES

NOTES

NOTES

NOTES

NOTES

NOTES

NOTES